Manual of
Individual Water
Supply Systems

U.S. Environmental Protection Agency
Office of Drinking Water

Fredonia Books
Amsterdam, The Netherlands

Manual of Individual Water Supply Systems

by
U. S. Environmental Protection Agency Office of
Drinking Water

ISBN: 1-58963-407-1

Copyright © 2001 by Fredonia Books

Reprinted from the 1982 edition

Fredonia Books
Amsterdam, the Netherlands
http://www.fredoniabooks.com

Acknowledgment

This manual follows in general the format of its predecessor, Public Health Service Publication No. 24, prepared by the Joint Committee on Rural Sanitation.[1] The Water Supply Division is indebted to that committee for the many important contributions that have been retained.

The special committee charged with responsibility for preparing this new manual was composed of the following persons:

W. J. Whitsell (Committee Chairman), Ground Water Engineer, Water Supply Division, Environmental Protection Agency

R. D. Lee, Chief, Surveillance and Technical Assistance, Water Supply Division, Environmental Protection Agency

E. L. Hockman, Ground Water Engineer, Water Supply Division, Environmental Protection Agency

D. K. Keech, Chief, Ground Water Quality Control Section, Michigan State Department of Health

G. F. Briggs, Vice President—Engineering, U. O. P. Johnson Division, St. Paul, Minn.

Ed Norman, Marvin Norman Drilling Co., Vienna, W. Va.

It is impractical to list here all persons and organizations that have offered valuable criticisms and suggestions for improvements. Some 30 Federal, State, and private agencies participated in the final technical review of the completed draft. Their contributions led to a considerable number of improvements. To all of them, the Water Supply Division expresses its sincere gratitude.

James H. McDermott, Director
Water Supply Division

[1] The Joint Committee on Rural Sanitation was composed of specialists from the following agencies U.S. Department of Agriculture, U.S. Department of Health, Education, and Welfare, U.S. Department of the Interior, U.S. Coast Guard, Federal Housing Administration, Veterans' Administration, Tennessee Valley Authority, Conference of State Sanitary Engineers, Water Pollution Control Federation, Conference of Municipal, Public Health Engineers, American Public Health Association, National Water Well Association, and Water Systems Council.

Foreword

Healthful, comfortable living requires the availability of an adequate supply of good quality water for drinking and domestic purposes.

Whenever feasible, the consumer will do well to obtain his water from a public water system in order to enjoy the advantages of qualified supervision under the control of a responsible public agency. It is usually his best assurance of an uninterrupted supply of safe water.

It is not always possible, or economically feasible, to obtain water from a community water system, and the consumer is then faced with the need to choose an alternative supply. It is to the individual or institution faced with this need that this manual is primarily directed.

This manual is a revision of PHS Publication No. 24, *Individual Water Supply Systems,* published in 1962. The revision was begun by the Bureau of Water Hygiene of the U.S. Public Health Service in 1969. In late 1970, the Bureau's activities and personnel were transferred by law to the newly created Environmental Protection Agency (EPA). Work on the manual was completed within EPA.

The Water Supply Division hopes the manual will be useful to Federal agencies concerned with the development of individual water supplies, and to State and local health departments, well drillers, contractors, and individual homeowners as well as to owners and operators of private and public institutions.

Persons familiar with PHS Publication No. 24 will note that extensive rewriting and expansion of certain portions have taken place. This is especially notable in the sections dealing with ground water and wells. The changes reflect primarily the experiences of the past 10 years, and especially the advent of new equipment, methods, and materials. Totally new sections, new illustrations, and new tables have been added to provide more complete coverage of the subjects. Color has been used to clarify illustrations. Particular attention has been paid to the need for keeping recommendations on construction as practical as possible without compromising quality and basic principles of sanitation.

Since a considerable portion of the manual deals with ground water and wells — over 90 percent of individual systems — the special committee organized to assemble the new manual was reinforced appropriately with persons who have had extensive practical experience in water well construction. Their contributions were in turn reviewed by individuals and organizations whose work keeps them in close contact with the field application of practices recommended in this manual.

Changing times and changing living habits have imposed greater and greater pollution loads on our environment. It is imperative that all water systems be constructed in accordance with recommended practices known to provide effective defenses against contamination. In addition, each recommendation has been carefully studied to make sure that it meets the following other important requirements:

1. It must be practical, yielding results with equipment and techniques currently available.
2. Its cost must be consistent with the benefits to be expected from its execution.
3. It must make an important contribution to the useful life of the installation.

Assistance in the planning of individual water systems usually can be obtained from State or local health departments. The health department may in turn suggest other agencies — health departments and departments of geology and water resources. These agencies should be the first contacts.

If any question of water rights is involved, the owner should seek legal advice. Departments of geology and water resources can frequently advise as to whether water rights are likely to be a problem.

Contents

List of Tables

List of Illustrations

Part I

Selection of a Water Source

The planning of an individual water supply system requires a determination of the quality of the water and available sources. In addition, it is desirable for one to have a basic knowledge of water rights and the hydrological, geological, chemical, biological, and possible radiological factors affecting the water. These factors are usually interrelated because of the continuous circulation of the water or water vapor from the oceans to the air, over the surface of the land and underground, and back to the oceans. This circulation is called hydrologic cycle. (See fig. 1.)

RIGHTS TO THE USE OF WATER

The right of an individual to use of water for domestic, irrigation, or other purposes varies in different States. Some water rights stem from ownership of the land bordering or overlying the source, while others are acquired by a performance of certain acts required by law.

There are three basic types of water rights. They are:

Riparian.—Rights that are acquired together with title to the land bordering or overlying the source of water.

Appropriative —Rights that are acquired by following a specific legal procedure.

Prescriptive —Rights that are acquired by diverting and putting to use, for a period specified by statute, water to which other parties may or may not have prior claims. The procedure necessary to obtain prescriptive rights must conform with the conditions established by the water rights laws of individual States.

When there is any question regarding the right to the use of water, the property owner should consult the appropriate authority in his State and clearly establish his rights to its use.

SOURCES OF WATER SUPPLY

At some time in its history, water resided in the oceans. By evaporation, moisture is transferred from the ocean surface to the atmosphere, where the winds carry the moisture-laden air over landmasses. Under certain conditions, this water vapor condenses to form clouds, which release their moisture as precipitation in the form of rain, hail, sleet, or snow.

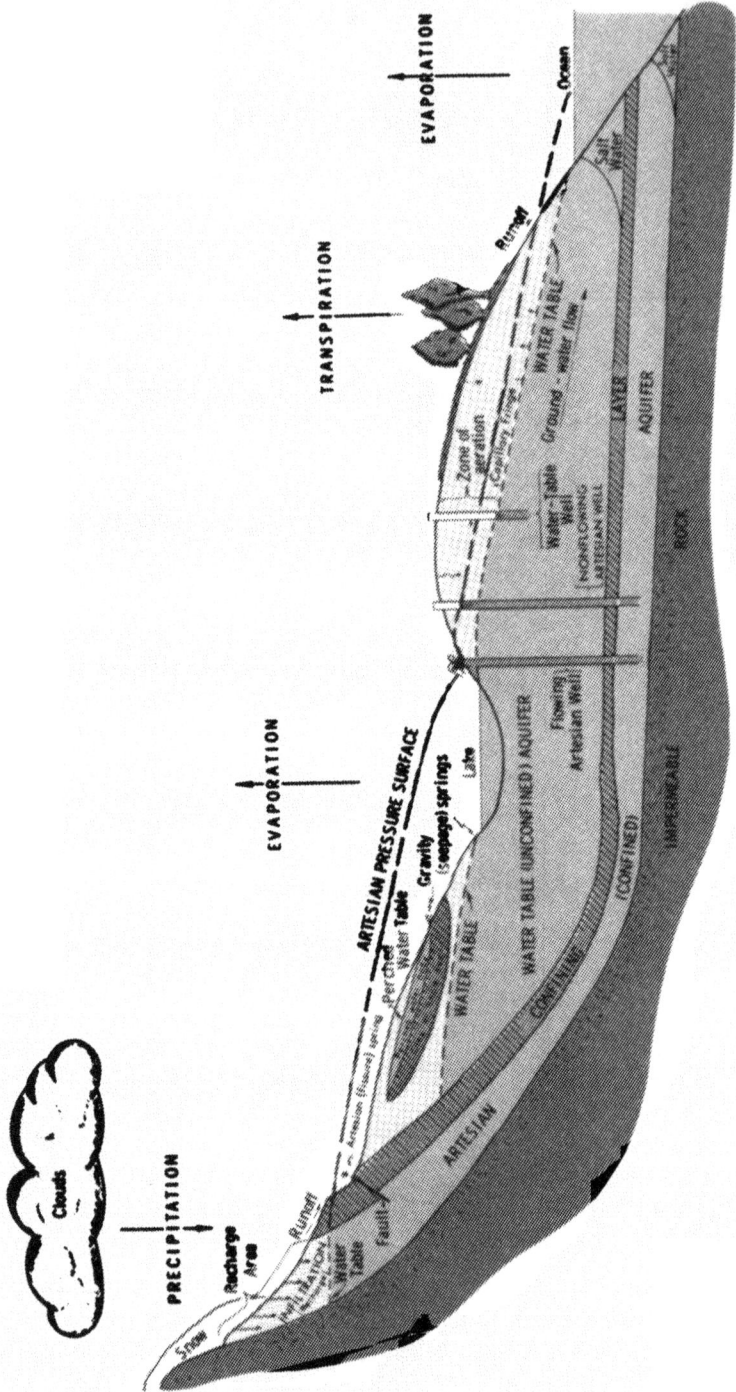

FIGURE 1. The hydrologic cycle.

When rain falls toward the earth, a part may evaporate and return immediately to the atmosphere. Precipitation in excess of the amount that wets a surface or supplies evaporation requirements is available as a potential source of water supply.

Ground Water

A part of the precipitation may infiltrate into the soil. (See fig. 1.) This water replenishes the soil moisture or is used by growing plants and returned to the atmosphere by transpiration. Water that drains downward below the root zone finally reaches a level at which all of the openings or voids in the earth's materials are filled with water. This zone is known as the "zone of saturation." Water in the zone of saturation is referred to as "ground water." The upper surface of the zone of saturation, if not confined by impermeable material, is called the "water table." When an overlying impermeable formation confines the water in the zone of saturation under a pressure greater than atmospheric pressure, the ground water is under artesian pressure. The name "artesian" comes from the ancient province of Artesium in France, where in the days of the Romans water flowed to the surface of the ground from a well. Not all water from wells that penetrate artesian formations flows above the surface of the land. For a well to be artesian, the water in the well must stand above the top of the aquifer. An aquifer, or water-bearing formation, is an underground layer of permeable rock or soil that permits the passage of water.

The porous material just above the water table may contain water by capillarity in the smaller void spaces. This zone is referred to as the "capillary fringe." It is not a source of supply since the water held will not drain freely by gravity.

Because of the irregularities in underground deposits and in surface topography, the water table occasionally intersects the surface of the ground or the bed of a stream, lake, or ocean. As a result, ground water moves to these locations and out of the aquifer or ground water reservoir. Thus, ground water is continually moving within the aquifer even though the movement may be slow. The water table or artesian pressure surface slopes from areas of recharge to areas of discharge. The pressure differences represented by these slopes cause the flow of ground water within the aquifer. At any point the slope is a reflection of the rate of flow and resistance to movement of water through the saturated formation. Seasonal variations in the supply of water to the underground reservoir cause considerable changes in the elevation and slope of the water table and artesian pressure level.

Wells

A well can be used to extract water from the ground water reservoir. Pumping will cause a lowering of the water table near the well. If pumping continues at a rate that exceeds the rate at which the water may be replaced by the water-bearing formations, the sustained yield of the well is exceeded. If wells extract water from an aquifer over a period of time at a rate such that the aquifer will become depleted or bring about other undesired results, then the "safe yield" of the aquifer is exceeded. Under these conditions, salt-water encroachment may occur where wells are located near the seashore or other surface or underground saline waters.

Springs

An opening in the ground surface from which ground water flows is a spring. Water may flow by force of gravity (from water-table aquifers), or be forced out by artesian pressure. The flow from a spring may vary considerably. When the water-table or artesian pressure fluctuates, so does the flow of springs. For further discussion, see part II.

Surface Water

Precipitation that does not enter the ground through infiltration or is not returned to the atmosphere by evaporation flows over the ground surface and is classified as direct runoff. Direct runoff is water that moves over saturated or impermeable surfaces, and in stream channels or other natural or artificial storage sites. The dry weather (base) flow of streams is derived from ground water or snowmelt.

In some areas, a source of water for individual development is the rainfall intercepted by roof surfaces on homes, barns, or other buildings. Water from such impermeable surfaces can be collected and stored in tanks called cisterns. In some instances, natural ground surfaces can be conditioned to make them impermeable. This conditioning will increase runoff to cisterns or large artificial storage reservoirs, thereby reducing loss by infiltration into the ground.

Runoff from ground surfaces may be collected in either natural or artificial reservoirs. A portion of the water stored in surface reservoirs is lost by evaporation and from infiltration to the ground water table from the pond bottom. Transpiration from vegetation in and adjacent to ponds constitutes another means of water loss.

Ground and Surface Water

Ground water may become surface water at springs or at intersections of a water body and a water table. During extended dry periods, stream flows consist largely of water from the ground water reservoir. As the ground water reservoir is drained by the surface stream, the flow will reach a minimum or may cease altogether. It is important in evaluating stream and spring supplies to consider seasonal fluctuations in flow.

Snow

Much of the snow falling on a water shed is kept in storage on the ground surface until temperatures rise above freezing. In the mountainous areas of the western United States, snow storage is an important source of water supply through much of the normal irrigation season. Measures taken to increase the snowpack and reduce the melt rate are usually beneficial to individual water supply systems in these areas.

QUALITY OF WATER

Precipitation in the form of rain, snow, hail, or sleet contains very few impurities. It may contain trace amounts of mineral matter, gases, and other substances as it forms and falls through the earth's atmosphere. The precipitation, however, has virtually no bacterial content.

Once precipitation reaches the earth's surface, many opportunities are presented for the introduction of mineral and organic substances, micro-organisms, and other forms of pollution (contamination).[1] When water runs over or through the ground surface, it may pick up particles of soil. This is noticeable in the water as cloudiness or turbidity. It also picks up particles of organic matter and bacteria. As surface water seeps downward into the soil and through the underlying material to the water table, most of the suspended particles are filtered out. This natural filtration may be partially effective in removing bacteria and other particulate materials; however, the chemical characteristics of the water may change and vary widely when it comes in contact with mineral deposits. Chemical and bacteriological analyses may be performed by a State or local health department or by a commercial laboratory.

The widespread use of synthetically produced chemical compounds, including pesticides and insecticides, has caused a renewed interest in the quality of water. Many of these materials are known to be toxic and others have certain undesirable characteristics which interfere with the use of the water even when present in relatively small concentrations. In recent years instances of water pollution have been traced to a sewage or waste water source containing synthetic detergents.

Substances that alter the quality of water as it moves over or below the surface of the earth may be classified under four major headings.

---

[1] Pollution as used in this manual means the presence in water of any foreign substances (organic, inorganic, radiological, or biological) which tend to lower its quality to a point that it constitutes a health hazard or impairs the usefulness of the water. Contamination, where used in this manual, has essentially the same meaning

1. *Physical* Physical characteristics relate to the quality of water for domestic use and are usually associated with the appearance of water, its color or turbidity, temperature, taste, and odor in particular.
2. *Chemical.* Chemical differences between waters are sometimes evidenced by their observed reactions, such as the comparative performance of hard and soft waters in laundering.
3. *Biological.* Biological agents are very important in their relation to public health and may also be significant in modifying the physical and chemical characteristics of water.
4. *Radiological* Radiological factors must be considered in areas where there is a possibility that the water may have come in contact with radioactive substances.

Consequently, in the development of an individual water supply system, it is necessary to examine carefully all the factors that might adversely affect the intended use of a water supply source.

Physical Characteristics

The water as used should be free from all impurities that are offensive to the sense of sight, taste, or smell. The physical characteristics of the water include turbidity, color, taste and odor, temperature, and foamability.

Turbidity The presence of suspended material such as clay, silt, finely divided organic material, plankton, and other inorganic material in water is known as turbidity. Turbidities in excess of 5 units are easily detectable in a glass of water, and are usually objectionable for esthetic reasons.

Clay or other inert suspended particles in drinking water may not adversely affect health, but water containing such particles may require treatment to make it suitable for its intended use. Following a rainfall, variations in the ground water turbidity may be considered an indication of surface or other introduced pollution.

Color. Dissolved organic material from decaying vegetation and certain inorganic matter cause color in water. Occasionally, excessive blooms of algae or the growth of aquatic micro-organisms may also impart color. While color itself is not usually objectionable from the standpoint of health, its presence is esthetically objectionable and suggests that the water needs appropriate treatment.

Taste and Odor. Taste and odor in water can be caused by foreign matter such as organic compounds, inorganic salts, or dissolved gases. These materials may come from domestic, agricultural, or natural sources. Acceptable waters should be free from any objectionable taste or odor at point of use. Knowledge concerning the chemical quality of a water supply source is important in order to determine what treatment, if any, is required to make the water acceptable for domestic use.

Temperature. The most desirable drinking waters are consistently cool and do not have temperature fluctuations of more than a few degrees. Ground water and surface water from mountainous areas generally meet these criteria. Most individuals find that water having a temperature between 50° and 60°F is most palatable.

Foamability. Since 1965 the detergent formulations have been changed to eliminate alkyl benzene sulfonate (ABS), which was very slowly degraded by nature. The more rapidly biodegradable linear alkylate sulfonate (LAS) has been substituted in most detergents. Even LAS is not degraded very rapidly in the absence of oxygen — a condition that exists in cesspools and some septic tank tile fields.

Foam in water is usually caused by concentrations of detergents greater than 1 milligram per liter. While foam itself is not hazardous, the user should understand that if enough detergent is reaching a water supply to cause a noticeable froth to appear on a glass of water, other possibly hazardous materials of sewage origin are also likely to be present.

Chemical Characteristics

The nature of the rocks that form the earth's crust affects not only the quantity of water that may be recovered but also its characteristics. As surface water seeps downward to the water table, it dissolves portions of the minerals contained by soils and rocks. Ground water, therefore, usually contains more dissolved minerals than surface water.

The chemical characteristics of water in a particular locality can sometimes be predicted from analyses of adjacent water sources. These data are often available in published reports of the U.S. Geological Survey or from Federal, State, and local health, geological, and water agencies. In the event that the information is not available, a chemical analysis of the water source should be made. Some State health and geological departments, as well as State colleges, and many commercial laboratories have the facilities and may be able to provide this service.

Information that can be obtained from a chemical analysis is —

- The possible presence of harmful or disagreeable substances
- The potential for the water to corrode parts of the water system
- The tendency for the water to stain fixtures and clothing

The size of sample required and the method of collection should be in accordance with recommendations of the facility making the analysis.

The following is a discussion of the chemical characteristics of water based on the limits recommended by the U.S. Environmental Protection Agency.[2]

Toxic Substances. Water may contain toxic substances in solution. If analysis of the water supply shows that these substances exceed the following concentrations, the supply should *not* be used

Substance	Milligrams per liter[1]	Substance	Milligrams per liter 1*
Arsenic (As)	0 05	Fluoride (F)	(2*)
Barium (Ba) .	1 00	Lead (Pb) .	0 05
Cadmium (Cd) . .	01	Mercury . .	0 002
Chromium (Cr^{+6})	05	Selenium (Se)	01
		Silver (Ag) .	05

1* The term "milligrams per liter (mg/ℓ)" replaces the term "parts per million (ppm) " For water, the two terms are essentially equivalent
2* See following table.

The maximum concentrations of fluoride depend on the annual average maximum daily air temperature, as shown in the following table, because the temperature influences water intake.

Annual average of maximum daily air temperature	Maximum allowable fluoride concentration (mg/ℓ)
50 0°-53 7° F	2 4
53 8°-58 3° F	2 2
58 4°-63 8° F	2 0
63 9°-70 6° F	1 8
70 7°-79 2° F	1.6
79 3°-90 5° F	1 4

Chlorides Most waters contain some chloride in solution. The amount present can be caused by the leaching of marine sedimentary deposits, by pollution from sea water, brine, or industrial and domestic wastes. Chloride concentrations in excess of about 250 mg/ℓ usually produce a noticeable taste in drinking water. In areas where the chloride content is higher than 250 mg/ℓ and all other criteria are met, it may be necessary to use a water source that exceeds this limit.

An increase in chloride content in water may indicate possible pollution from sewage sources, particularly if the normal chloride content is known to be low.

---

[2] U S Environmental Protection Agency, Office of Drinking Water, "National Interim Primary Drinking Water Regulations," December 1975 U S Environmental Protection Agency, Washington, D C 20460

Copper. Copper is found in some natural waters, particularly in areas where these ore deposits have been mined.

Excessive amounts of copper can occur in corrosive water that passes through copper pipes. Copper in small amounts is not considered detrimental to health, but will impart an undesirable taste to the drinking water. For this reason, the recommended limit for copper is 1.0 mg/ℓ.

Fluorides. In some areas water sources contain natural fluorides. Where the concentrations approach optimum levels, beneficial health effects have been observed. In such areas the incidence of dental caries has been found to be below the rate in areas without natural fluorides.[4] The optimum fluoride level for a given area depends upon air temperature, since that is what primarily influences the amount of water people drink. Optimum concentrations from 0.7 to 1.2 mg/ℓ are recommended. Excessive fluorides in drinking water supplies may produce fluorosis (mottling) of teeth, which increases as the optimum fluoride level is exceeded. The State or local health departments, therefore, should be consulted for their recommendations.

Iron. Small amounts of iron are frequently present in water because of the large amount of iron present in the soil and because corrosive water will pick up iron from pipes. The presence of iron in water is considered objectionable because it imparts a brownish color to laundered goods and affects the taste of beverages such as tea and coffee. Recent studies indicate that eggs spoil faster when washed in water containing iron in excess of 10 mg/ℓ. The recommended limit for iron is 0.3 mg/ℓ.

Lead A brief or prolonged exposure of the body to lead can be seriously injurious to health. Prolonged exposure to relatively small quantities may result in serious illness or death. Lead taken into the body in quantities in excess of certain relatively low "normal" limits is a cumulative poison. A maximum concentration of 0.05 mg/ℓ of lead in water must not be exceeded. Excessive lead may occur in the source water, but the usual cause of excessive lead is corrosive water in contact with lead-painted roofs or the use of lead pipes. These conditions must be corrected to provide a safe water supply.

Manganese. There are two reasons for limiting the concentration of manganese in drinking water: (1) to prevent esthetic and economic damage, and (2) to avoid any possible physiological effects from excessive intake. The domestic user finds that manganese produces a brownish color in laundered goods, and impairs the taste of beverages, including coffee and tea. The recommended limit for manganese is 0.05 mg/ℓ.

---

[4]It is a known fact that the addition of about 1 mg/ℓ of fluoride to water supplies will help to prevent tooth decay in children. Some natural water supplies already contain amounts of fluoride that exceed the recommended optimum concentrations.

Nitrates Nitrate (NO_3) has caused methemoglobinemia (infant cyanosis or "blue baby disease") in infants who have been given water or fed formulas prepared with water having high nitrates. A domestic water supply should not contain nitrate concentrations in excess of 45 mg/ℓ (10 mg/ℓ expressed as nitrogen). Nitrates in excess of normal concentrations, often in shallow wells, may be an indication of seepage from livestock manure deposits. In some polluted wells, nitrite will also be present in concentrations greater than 1 mg/ℓ and is even more hazardous to infants. When the presence of high nitrite concentration is suspected the water should not be used for infant feeding. The nitrate concentration should be determined, and if excessive, advice should be obtained from health authorities about the suitability of using the water for drinking by anyone.

Pesticides. Careless use of pesticides can contaminate water sources and make the water unsuitable for drinking. Numerous cases have been reported where individual wells have been contaminated when the house was treated for termite control. The use of pesticides near wells is not recommended.

Sodium. When it is necessary to know the precise amount of sodium present in a water supply, a laboratory analysis should be made. When home water softeners utilizing the ion-exchange method are used, the amount of sodium will be increased. For this reason, water that has been softened should be analyzed for sodium when a precise record of individual sodium intake is recommended.

For healthy persons, the sodium content of water is unimportant because the intake from salt is so much greater, but for persons placed on a low-sodium diet because of heart, kidney, or circulatory ailments or complications of pregnancy, sodium in water must be considered. The usual low-sodium diets allow for 20 mg/ℓ sodium in the drinking water. When this limit is exceeded, such persons should seek a physician's advice on diet and sodium intake.

Sulfates. Waters containing high concentrations of sulfate caused by the leaching of natural deposits of magnesium sulfate (Epsom salts) or sodium sulfate (Glauber's salt) may be undesirable because of their laxative effects. Sulfate content should not exceed 250 mg/ℓ.

Zinc. Zinc is found in some natural waters, particularly in areas where these ore deposits have been mined. Zinc is not considered detrimental to health, but it will impart an undesirable taste to drinking water. For this reason, the recommended limit for zinc is 5.0 mg/ℓ.

Serious surface and ground water pollution problems have developed from existing and abandoned mining operations. Among the worst are those associated with coalmine operations, where heavy concentrations of iron, manganese, sulfates, and acids have resulted from the weathering and leaching of minerals (pyrites).

Chemical Terms

Alkalinity. Alkalinity is imparted to water by bicarbonate, carbonate, or hydroxide components. The presence of these compounds is determined by standard methods involving titration with various indicator solutions. Knowledge of the alkalinity components is useful in the treatment of water supplies.

Hardness. Hard water and soft water are relative terms. Hard water retards the cleaning action of soaps and detergents, causing an expense in the form of extra work and cleaning agents. Furthermore, when hard water is heated it will deposit a hard scale (as in a kettle, heating coils, or cooking utensils) with a consequent waste of fuel.

Calcium and magnesium salts, which cause hardness in water supplies, are divided into two general classifications: carbonate or temporary hardness and noncarbonate or permanent hardness.

Carbonate or temporary hardness is so called because heating the water will largely remove it. When the water is heated, bicarbonates break down into insoluble carbonates that precipitate as solid particles which adhere to a heated surface and the inside of pipes.

Noncarbonate or permanent hardness is so called because it is not removed when water is heated. Noncarbonate hardness is due largely to the presence of the sulfates and chlorides of calcium and magnesium in the water.

pH. pH is a measure of the hydrogen ion concentration in water. It is also a measure of the acid or alkaline content. pH values range from 0 to 14, where 7 indicates neutral water; values less than 7, increasing acidity; and values greater than 7, increasing alkalinity. The pH of water in its natural state often varies from 5.5 to 9.0. Determination of the pH value assists in the control of corrosion, the determination of proper chemical dosages, and adequate control of disinfection.

Biological Factors

Water for drinking and cooking purposes must be made free from disease-producing organisms. These organisms include bacteria, protozoa, virus, and helminths (worms).

Contamination of Water Supplies

Some organisms that cause disease in man originate with the fecal discharges of infected individuals. It is seldom practical to monitor and control the activities of human disease carriers. For this reason, it is necessary to exercise precautions against contamination of a normally safe water source or to institute treatment methods which will produce a safe water.

Unfortunately, the specific disease-producing organisms present in water are not easily identified. The techniques for comprehensive bacteriological examination are complex and time consuming. It has been necessary to develop tests that indicate the relative degree of contamination in terms of an easily defined quantity. The most widely used test involves estimation of the number of bacteria of the coliform group, which is always present in fecal wastes and outnumbers disease-producing organisms. The coliform group normally inhabits the intestinal tract of man, but is also found in most domestic animals and birds, as well as certain wild species.

Bacteriological Quality

The Public Health Service Drinking Water Standards have established limits for the mean concentration of coliform bacteria in a series of water samples and the frequency at which concentrations may exceed the mean. The results are expressed either in terms of a direct count of bacteria per unit volume – if the membrane filter (MF) procedure is used – or in terms of the "most probable number" (MPN). This latter term is an estimate based on mathematical formulas of probability.

The recommended standards for drinking water are roughly equivalent to restricting the coliform concentration to not more than one organism for each 100 milliliters of water.[5]

Application of the Public Health Service Drinking Water Standards to individual water supplies is difficult due to the low frequency with which samples can be properly collected and examined. Bacteriological examinations indicate the presence or absence of contamination in the collected sample only, and are indicative of quality only at the time of collection. A sample positive for coliforms is a good indication that the source may have been contaminated by surface washings or fecal material. *On the other hand, a negative result cannot be considered assurance of a continuously safe supply unless the results of a thorough sanitary survey of the surrounding area, together with subsequent negative samples, support this position.*

Collection of Samples for Bacteriological Examination

For a reliable indication of the bacteriological safety of an individual water supply, the owner should depend on the experience of qualified public health personnel. Special precautions are necessary in the collection of water samples, and proper training and experience are essential in evaluating the analytical results. Before a sample is collected, the examining facility should be contacted to obtain its recommendations. In the event that a procedure is not given, one should follow the suggestions found in appendix B.

---

[5] One hundred milliliters is about one-half cup in volume.

Other Biological Factors

Certain forms of aquatic vegetation and microscopic animal life in natural water may be either stimulated or retarded in their growth cycles by physical, chemical, or biological factors. For example, the growth of algae, minute green plants usually found floating in surface water, is stimulated by light, heat, nutrients such as nitrogen and phosphorus, and the presence of carbon dioxide as a product of organic decomposition. Their growth may, in turn, be retarded by changes in pH (measure of acidity), the presence of inorganic impurities, excessive cloudiness or darkness, temperature, and the presence of certain bacterial species.

Continuous cycles of growth and decay of algal cell material may result in the production of noxious byproducts that may adversely affect the quality of a water supply. The same general statements may be made regarding the growth cycles of certain nonpathogenic bacteria or microcrustacea that inhabit natural waters.

A water source should be as free from biological activity as possible. Biological activity can be avoided or kept to a minimum by:

1. Selecting water sources that do not normally support much plant or animal life.
2. Protecting the supply against subsequent contamination by biological agents.
3. Minimizing entrance of fertilizing materials, such as organic and nutrient minerals.
4. Controlling the light and temperature of stored water.
5. Providing treatment for the destruction of biologic life or its byproducts.

Radiological Factors

The development and use of atomic energy as a power source and mining of radioactive materials have made it necessary to establish limiting concentrations for the intake into the body of radioactive substances, including drinking water.

The effects of human exposure to radiation or radioactive materials are viewed as harmful and any unnecessary exposure should be avoided. The concentrations of radioactive materials specified in the current Public Health Service Drinking Water Standards are intended to limit the human intake of these substances so that the total radiation exposure of any individual will not exceed those defined in the Radiation Protection Guides recommended by the Federal Radiation Council. Man has always been exposed to natural radiation from water, food, and air. The amount of radiation to which the individual is normally exposed varies with the amount of background radioactivity. Water of high radioactivity is unusual. Nevertheless it is known to exist in certain areas, either from natural or manmade sources.

Radiological data indicating both background and other forms of radioactivity in an area may be available in publications of the U.S. Environmental Protection Agency, U.S. Public Health Service, U.S. Geological Survey, or from Federal, State, or local agencies. For information or recommendations on specific problems, the appropriate agency should be contacted.

QUANTITY OF WATER

One of the first steps in the selection of a suitable water supply source is determining the demand which will be placed on it. The essential elements of water demand include the average daily water consumption and the peak rate of demand. The average daily water consumption must be estimated—

1. To determine the ability of the water source to meet continuing demands over critical periods when surface flows are low and ground water tables are at minimum elevations and
2. For purposes of estimating quantities of stored water that would sustain demands during these critical periods.

The peak demand rates must be estimated in order to determine plumbing and pipe sizing, pressure losses, and storage requirements necessary to supply sufficient water during periods of peak water demand.

Average Daily Water Use

Many factors influence water use for a given system. For example, the mere fact that water under pressure is available stimulates its use for watering lawns and gardens, for washing automobiles, for operating air-conditioning equipment, and for performing many other utility activities at home and on the farm Modern kitchen and laundry appliances, such as food waste disposers and automatic dishwashers, contribute to a higher total water use and tend to increase peak demands. Since water requirements will influence all features of an individual development or improvement, they must figure prominently in plan preparation. Table1 presents a summary of average water use as a guide in preparing estimates, with local adaptations where necessary.

Peak Demands

The rate of water use for an individual water system will vary directly with domestic activity in the home or with the operational farm program. Rates are generally highest in the home near mealtimes, during midmorning laundry periods, and shortly before bedtime. During the intervening daytime hours and at night, water use may be virtually nil. Thus, the total amount of water used by a household may be distributed over only a few hours of the day,

TABLE 1. — *Planning guide for water use*

Types of establishments	Gallons per day
Airports (per passenger)	3-5
Apartments, multiple family (per resident)	60
Bath houses (per bather)	10
Camps.	
Construction, semipermanent (per worker)	50
Day with no meals served (per camper)	15
Luxury (per camper)	100-150
Resorts, day and night, with limited plumbing (per camper)	50
Tourist with central bath and toilet facilities (per person)	35
Cottages with seasonal occupancy (per resident)	50
Courts, tourist with individual bath units (per person)	50
Clubs.	
Country (per resident member)	100
Country (per nonresident member present)	25
Dwellings.	
Boardinghouses (per boarder)	50
Additional kitchen requirements for nonresident boarders	10
Luxury (per person)	100-150
Multiple-family apartments (per resident)	40
Rooming houses (per resident)	60
Single family (per resident)	50-75
Estates (per resident)	100-150
Factories (gallons per person per shift)	15-35
Highway rest area (per person)	5
Hotels with private baths (2 persons per room)	60
Hotels without private baths (per person)	50
Institutions other than hospitals (per person)	75-125
Hospitals (per bed)	250-400
Laundries, self-serviced (gallons per washing, i.e., per customer)	50
Livestock (per animal)	
Cattle (drinking)	12
Dairy (drinking and servicing)	35
Goat (drinking)	2
Hog (drinking)	4
Horse (drinking)	12
Mule (drinking)	12
Sheep (drinking)	2
Steer (drinking)	12
Motels with bath, toilet, and kitchen facilities (per bed space)	50
With bed and toilet (per bed space)	40
Parks:	
Overnight with flush toilets (per camper)	25
Trailers with individual bath units, no sewer connection (per trailer)	25
Trailers with individual baths, connected to sewer (per person)	50
Picnic:	
With bathhouses, showers, and flush toilets (per picnicker)	20
With toilet facilities only (gallons per picnicker)	10
Poultry·	
Chickens (per 100)	5-10
Turkeys (per 100)	10-18

TABLE 1. – *Planning guide for water use – Continued*

Types of establishments	Gallons per day
Restaurants with toilet facilities (per patron)	7-10
Without toilet facilities (per patron) ·	2½-3
With bars and cocktail lounge (additional quantity per patron) . . .	2
Schools	
Boarding (per pupil)	75-100
Day with cafeteria, gymnasiums, and showers (per pupil) . .	25
Day with cafeteria but no gymnasiums or showers (per pupil) .	20
Day without cafeteria, gymnasiums, or showers (per pupil) . .	15
Service stations (per vehicle)	10
Stores (per toilet room)	400
Swimming pools (per swimmer)	10
Theaters	
Drive-in (per car space)	5
Movie (per auditorium seat)	5
Workers	
Construction (per person per shift)	50
Day (school or offices per person per shift) . . .	15

during which the actual use is much greater than the average rate determined from Table 1.

Simultaneous operation of several plumbing fixtures will determine the maximum peak rate of water delivery for the home water system. For example, a shower, an automatic dishwasher, a lawn-sprinkler system, and a flush valve toilet all operated at the same time would probably produce a near-critical peak. It is true that not all of these facilities are usually operated together; but if they exist on the same system, there is always a possibility that a critical combination may result, and for design purposes this method of calculation is sound. Table 2 summarizes the rate of flow which would be expected for certain household and farm fixtures.

Special Water Considerations

Lawn Sprinkling. The amount of water required for lawn sprinkling depends upon the size of the lawn, type of sprinkling equipment, climate, soil, and water control. In dry or arid areas the amount of water required may equal or exceed the total used for domestic or farmstead needs. For estimating purposes, a rate of approximately ½ inch per hour of surface area is reasonable. This amount of water can be applied by sprinkling 30 gallons of water per hour over each 100 square feet.

Example

$$\frac{1000}{100} \times 30 = 300 \text{ gallons per hour or 5 gpm}$$

A lawn of 1,000 square feet would require 300 gallons per hour.

TABLE 2. – *Rates of flow for certain plumbing, household, and farm fixtures*

Location	Flow pressure[1] –pounds per square inch (psi)	Flow rate– gallons per minute (gpm)
Ordinary basin faucet	8	2.0
Self-closing basin faucet	8	2.5
Sink faucet, 3/8 inch	8	4.5
Sink faucet, 1/2 inch	8	4 5
Bathtub faucet	8	6.0
Laundry tub faucet, 1/2 inch	8	5.0
Shower	8	5.0
Ball-cock for closet	8	3.0
Flush valve for closet	15	[2]15-40
Flushometer valve for urinal	15	15.0
Garden hose (50 ft., 3/4-inch sill cock)	30	5.0
Garden hose (50 ft., 5/8-inch outlet)	15	3.33
Drinking fountains	15	.75
Fire hose 1-1/2 inches, 1/2-inch nozzle	30	40.0

[1]Flow pressure is the pressure in the supply near the faucet or water outlet while the faucet or water outlet is wide open and flowing.

[2]Wide range due to variation in design and type of closet flush valves.

When possible, the water system should have a minimum capacity of 500-600 gallons per hour. A water system of this size may be able to operate satisfactorily during a peak demand. Peak flows can be estimated by adding lawn sprinkling to peak domestic flows but not to fire flows.

Fire Protection. In areas of individual water supply systems, effective firefighting depends upon the facilities provided by the property owner. The National Fire Protection Association has prepared a report which outlines and describes ways to utilize available water supplies.[6]

The most important factors in successful firefighting are early discovery and immediate action. For immediate protection, portable fire extinguishers are desirable. Such first-aid protection is designed only for the control of fires in the early stage; therefore, a water supply is desirable as a second line of defense.

The use of gravity water supplies for firefighting presents certain basic problems. These include (1) the construction of a dam, farm pond, or storage tank to hold the water until needed, and (2) the determination of the size of pipeline installed from the supply. The

[6]National Fire Protection Association, "Water Supply Systems for Rural Fire Protection," *National Fire Codes*, vol. 8 (Boston, 1969).

size of the pipe is dependent upon two factors: (1) the total fall or head from the point of supply to the point of use and (2) the length of pipeline required.

A properly constructed well tapping a good aquifer can be a dependable source for both domestic use and fire protection. If the well is to be relied upon for fire protection without supplemental storage, it should demonstrate, by a pumping test, minimum capacity of 8 to 10 gallons per minute continuously for a period of 2 hours during the driest time of the year.

A more dependable installation results when motor, controls, and powerlines are protected from fire. A high degree of protection is achieved when all electrical elements are located outside at the well, and there is a separate powerline bypassing other buildings.

There are numerous factors determining the amount of fire protection that should be built into a water system. Publications of the National Fire Protection Association[7] provide more information on this subject.

The smallest individual pressure systems commercially available provide about 210 gallons per hour (3½ gallons per minute). While this capacity will furnish a stream, through an ordinary garden hose, of some value in combating incipient fires or in wetting down adjacent buildings, it cannot be expected to be effective on a fire that has gained any headway. When such systems are already installed, connections and hose should be provided. When a new system is being planned or a replacement of equipment made, it is urged that a capacity of at least 500 gallons an hour (8-1/3 gallons per minute) be specified and the supply increased to meet this demand. If necessary, storage should be added. The additional cost for the larger unit necessary for fire protection is partially offset by the increased quantities of water available for other uses.

SANITARY SURVEY

The importance of a sanitary survey of water sources cannot be overemphasized. With a new supply, the sanitary survey should be made in conjunction with the collection of initial engineering data covering the development of a given source and its capacity to meet existing and future needs. The sanitary survey should include the detection of all health hazards and the assessment of their present and future importance. Persons trained and competent in public health engineering and the epidemiology of waterborne diseases should conduct the sanitary survey. In the case of an existing supply, the sanitary survey should be made at a frequency compatible with the control of the health hazards and the maintenance of a good sanitary quality.

[7]Ibid.

The information furnished by the sanitary survey is essential to complete interpretation of bacteriological and frequently the chemical data. This information should always accompany the laboratory findings. The following outline covers the essential factors which should be investigated or considered in a sanitary survey. Not all of the items are pertinent to any one supply and, in some cases, items not in the list would be important additions to the survey list.

Ground Water Supplies

a. Character of local geology; slope of ground surface.

b. Nature of soil and underlying porous strata; whether clay, sand, gravel, rock (especially porous limestone); coarseness of sand or gravel; thickness of water-bearing stratum, depth to water table; location, log, and construction details of local wells in use and abandoned.

c. Slope of water table, preferably as determined from observational wells or as indicated, presumptively but not certainly, by slope of ground surface.

d. Extent of drainage area likely to contribute water to the supply.

e. Nature, distance, and direction of local sources of pollution.

f. Possibility of surface-drainage water entering the supply and of wells becoming flooded; methods of protection.

g. Methods used for protecting the supply against pollution by means of sewage treatment, waste disposal, and the like.

h. Well construction:
   1. Total depth of well.
   2. Casing: diameter, wall thickness, material, and length from surface.
   3. Screen or perforations: diameter, material, construction, locations, and lengths.
   4. Formation seal: material (cement, sand, bentonite, etc.), depth intervals, annular thickness, and method of placement.

i. Protection of well at top: presence of sanitary well seal, casing height above ground, floor, or flood level, protection of well vent, protection of well from erosion and animals.

j. Pumphouse construction (floors, drains, etc.), capacity of pumps, drawdown when pumps are in operation.

k. Availability of an unsafe supply, usable in place of normal supply, hence involving danger to the public health.

l. Disinfection: equipment, supervision, test kits, or other types of laboratory control.

Surface-Water Supplies

a. Nature of surface geology: character of soils and rocks.

b. Character of vegetation, forests, cultivated and irrigated land, including salinity, effect on irrigation water, etc.

c. Population and sewered population per square mile of catchment area.

d. Methods of sewage disposal, whether by diversion from watershed or by treatment.

e. Character and efficiency of sewage-treatment works on watershed.

f. Proximity of sources of fecal pollution to intake of water supply.

g. Proximity, sources, and character of industrial wastes, oil field brines, acid mine waters, etc.

h. Adequacy of supply as to quantity.

i. For lake or reservoir supplies: wind direction and velocity data, drift of pollution, sunshine data (algae).

j. Character and quality of raw water: coliform organisms (MPN), algae, turbidity, color, objectionable mineral constituents.

k. Nominal period of detention in reservoir or storage basin.

l. Probable minimum time required for water to flow from sources of pollution to reservoir and through reservoir intake.

m. Shape of reservoir, with reference to possible currents of water, induced by wind or reservoir discharge, from inlet to water-supply intake.

n. Protective measures in connection with the use of watershed to control fishing, boating, landing of airplanes, swimming, wading, ice cutting, permitting animals on marginal shore areas and in or upon the water, etc.

o. Efficiency and constancy of policing.

p. Treatment of water: kind and adequacy of equipment; duplication of parts; effectiveness of treatment; adequacy of supervision and testing; contact period after disinfection; free chlorine residuals carried.

q. Pumping facilities: pumphouse, pump capacity and standby units, storage facilities.

Part II

Ground Water

ROCK FORMATIONS AND THEIR WATER-BEARING PROPERTIES

The rocks that form the crust of the earth are divided into three classes:

1. *Igneous* Rocks that are derived from the hot magma deep in the earth. They include granite and other coarsely crystalline rocks, dense igneous rocks such as occur in dikes and sills, basalt and other lava rocks, cinders, tuff, and other fragmental volcanic materials.

2. *Sedimentary.* Rocks that consist of chemical precipitates and of rock fragments deposited by water, ice, or wind. They include deposits of gravel, sand, silt, clay, and the hardened equivalents of these – conglomerate, sandstone, siltstone, shale, limestone, and deposits of gypsum and salt.

3. *Metamorphic.* Rocks that are derived from both igneous and sedimentary rocks through considerable alteration by heat and pressure at great depths. They include gneiss, schist, quartzite, slate, and marble.

The pores, joints, and crevices of the rocks in the zone of saturation are generally filled with water. Although the openings in these rocks are usually small, the total amount of water that can be stored in the subsurface reservoirs of the rock formations is large. The most productive aquifers are deposits of clean, coarse sand and gravel; coarse, porous sandstones; cavernous limestones; and broken lava rock. Some limestones, however, are very dense and unproductive. Most of the igneous and metamorphic rocks are hard, dense, and of low permeability. They generally yield small quantities of water. Among the most unproductive formations are the silts and clays. The openings in these materials are too small to yield water, and the formations are structurally too incoherent to maintain large openings under pressure. Compact materials near the surface, with open joints similar to crevices in rock, may yield small amounts of water.

GROUND WATER BASINS

In an undeveloped ground water basin, movement of water to lower basins, seepage from and to surface-water sources, and transpiration are dependent upon the water in storage and the rate of recharge. During periods following abundant rainfall, recharge may exceed discharge. When recharge exceeds discharge, the excess rainfall increases the amount of water available in storage in the ground water basin. As the water table or artesian pressure rises, the gradients to points of discharge become steeper and outflows increase. When recharge ceases, storage decrease from outflow causes water-table levels and artesian pressures to decline. In most undeveloped basins the major fluctuations in storage are seasonal, with the mean annual elevation of water levels showing little variation. Thus, the average annual inflow to storage equals the average annual outflow, a quantity of water referred to as the basin yield.

The proper development of a ground water source requires careful consideration of the hydrological and geological conditions of the area. The individual who wishes to take full advantage of a water source for domestic use should obtain the assistance of a qualified ground water engineer, ground water geologist, hydrologist, or contractor familiar with the construction of wells in his area. He should rely on facts and experience, not on instinct or intuition. Facts on the geology and hydrology of an area may be available in publications of the U.S. Geological Survey or from other Federal and State agencies. The National Water Well Association[1] also offers assistance.

SANITARY QUALITY OF GROUND WATER

When water seeps downward through overlying material to the water table, particles in suspension, including micro-organisms, may be removed. The extent of removal depends on the thickness and character of the overlying material. Clay or "hardpan" provides the most effective natural protection of ground water. Silt and sand also provide good filtration if fine enough and in thick enough layers. The bacterial quality of the water also improves during storage in the aquifer because storage conditions are usually unfavorable for bacterial survival. Clarity alone does not guarantee that ground water is safe to drink; this can only be determined by laboratory testing.

Ground water found in unconsolidated formations (sand, clay, and gravel) and protected by similar materials from sources of pollution is more likely to be safe than water coming from consolidated formations (limestone, fractured rock, lava, etc.).

Where limited filtration is provided by overlying earth materials,

---

[1] 88 East Broad St., Columbus, Ohio 43215

water of better sanitary quality can sometimes be obtained by drilling deeper. It should be recognized, however, that there are areas where it is not possible, because of the geology, to find water at greater depths. Much unnecessary drilling has been done in the mistaken belief that more and better quality water can always be obtained by drilling to deeper formations.

In areas without central sewerage systems, human excreta are usually deposited in septic tanks, cesspools, or pit privies. Bacteria in the liquid effluents from such installations may enter shallow aquifers. Sewage effluents have been known to find their way directly into water-bearing formations by way of abandoned wells or soil-absorption systems. In such areas, the threat of contamination may be reduced by proper well construction, locating it farther from the source of contamination. The direction of ground water flow usually approximates that of the surface flow. It is always desirable to locate a well so that the normal movement of ground water flow carries the contaminant away from the well.

CHEMICAL AND PHYSICAL QUALITY OF GROUND WATER

The mineral content of ground water reflects its movement through the minerals which make up the earth's crust. Generally, ground water in arid regions is harder and more mineralized than water in regions of high annual rainfall. Also, deeper aquifers are more likely to contain higher concentrations of minerals in solution because the water has had more time (perhaps millions of years) to dissolve the mineral rocks. For any ground water region there is a depth below which salty water, or brine, is almost certain to be found. This depth varies from one region to another.

Some substances found naturally in ground water, while not necessarily harmful, may impart a disagreeable taste or undesirable property to the water. Magnesium sulfate (Epsom salt), sodium sulfate (Glauber's salt), and sodium chloride (common table salt) are but a few of these. Iron and manganese are commonly found in ground waters (see p. 9). It is an interesting fact that regular users of waters containing amounts of these substances considered by many to be excessive commonly become accustomed to the water and consider it to have a good taste!

Concentrations of chlorides and nitrates that are usually high for a particular region may be indicators of sewage pollution. This is another reason why a chemical analysis of the water (p. 7) should be made periodically and these results interpreted by someone familiar with the area.

TEMPERATURE

The temperature of ground water remains nearly constant throughout the year. Water from very shallow sources (less than 50 feet deep) may vary somewhat from one season to another, but water from deeper zones remains quite constant, its temperature being close to that for the average annual temperature at the surface. This is why water from a well may seem to be warm in winter or cold during the summer.

Contrary to popular opinion, colder water is *not* obtained by drilling deeper. Beyond about 100 feet of depth, the temperature of ground water increases steadily at the rate of about 1°F for each 75 to 150 feet of depth. In volcanic regions this rate of increase may be much greater.

DISTANCES TO SOURCES OF CONTAMINATION

All ground water sources should be located a safe distance from sources of contamination. In cases where sources are severely limited, however, a ground water aquifer that might become contaminated may be considered for a water supply if treatment is provided. After a decision has been made to locate a water source in an area, it is necessary to determine the distance the source should be placed from the origin of contamination and the direction of water movement. A determination of a safe distance is based on specific local factors described in the section on "Sanitary Survey" in part I of this manual.

Because many factors affect the determination of "safe" distances between ground water supplies and sources of pollution, it is impractical to set fixed distances. Where insufficient information is available to determine the "safe" distance, the distance should be the maximum that economics, land ownership, geology, and topography will permit. It should be noted that the direction of ground water flow does not always follow the slope of the land surface. Each installation should be inspected by a person with sufficient training and experience to evaluate all of the factors involved.

Since safety of a ground water source depends primarily on considerations of good well construction and geology, these factors should be the guides in determining safe distances for different situations. The following criteria apply only to properly constructed wells as described in this manual. There is no safe distance for a poorly constructed well!

When a properly constructed well penetrates an unconsolidated formation with good filtering properties, and when the aquifer itself is separated from sources of contamination by similar materials, research and experience have demonstrated that 50 feet is an adequate distance separating the two. Lesser distances should be accepted only after a comprehensive sanitary survey, conducted by qualified State or local health agency officials, has satisfied the officials that such lesser distances are both necessary and safe.

If it is proposed to install a properly constructed well in formations of unknown character, the State or U.S. Geological

Survey and the State or local health agency should be consulted.

When wells must be constructed in consolidated formations, extra care should always be taken in the location of the well and in setting "safe" distances, since pollutants have been known to travel great distances in such formations. The owner should request assistance from the State or local health agency.

The following table is offered as a guide in determining distances:

Formations	Minimum acceptable distance from well to source of contamination
Favorable (unconsolidated)..	50 feet. Lesser distances only on health department approval following comprehensive sanitary survey of proposed site and immediate surroundings.
Unknown	50 feet only after comprehensive geological survey of the site and its surroundings has established, to the satisfaction of the health agency, that favorable formations do exist.
Poor (consolidated) . .	Safe distances can be established only following both the comprehensive geological and comprehensive sanitary surveys. These surveys also permit determining the direction in which a well may be located with respect to sources of contamination. In no case should the acceptable distance be less than 50 feet.

EVALUATING CONTAMINATION THREATS TO WELLS

Conditions unfavorable to the control of contamination and that may require specifying *greater* distances between a well and sources of contamination are.

1. *Nature of the contaminant* Human and animal excreta and toxic chemical wastes are serious health hazards. Salts, detergents, and other substances that dissolve in water can mix with ground water and travel with it. They are not ordinarily removed by natural filtration.

2. *Deeper disposal* Cesspools, dry wells, disposal and waste injection wells, and deep leaching pits that reach aquifers or reduce the amount of filtering earth materials between the wastes and the aquifer increase the danger of contamination.

3. *Limited filtration.* When earth materials surrounding the well and overlying the aquifer are too coarse to provide effective filtration – as in limestone, coarse gravel, etc. – or when they form a layer too thin, the risk of contamination is increased.

4. *The aquifer* When the materials of the aquifer itself are too coarse to provide good filtration – as in limestone, fractured rock, etc. – contaminants entering the aquifer

through outcrops or excavations may travel great distances. It is especially important in such cases to know the direction of ground water flow and whether there are outcrops of the formation (or excavations reaching it) "upstream" and close enough to be a threat.

5. *Volume of waste discharged* Since greater volumes of wastes discharged and reaching an aquifer can significantly change the slope of the water table and the direction of ground water flow, it is obvious that heavier discharges can increase the threat of contamination.

6. *Contact surface.* When pits and channels are designed and constructed to increase the rate of absorption – as in septic tank leaching systems, cesspools, and leaching pits – more separation from the water source will be needed than when tight sewer lines or waste pipes are used.

7. *Concentration of contamination sources.* The existence of more than one source of contamination contributing to the general area increases the total pollution load and, consequently, the danger of contamination.

DEVELOPMENT OF GROUND WATER

The type of ground water development to be undertaken is dependent upon the geological formations and hydrological characteristics of the water-bearing formation. The development of ground water falls into two main categories:

    1. Development by wells
       a. Nonartesian or water table
       b. Artesian
    2. Development from springs
       a. Gravity
       b. Artesian

Nonartesian wells are those that penetrate formations in which ground water is found under water-table conditions. Pumping from the well lowers the water table in the vicinity of the well and water moves toward the well under the pressure differences thus artificially created.

Artesian wells are those that penetrate aquifers in which the ground water is found under hydrostatic pressure. Such a condition occurs in an aquifer that is confined beneath an impermeable layer of material at an elevation lower than that of the intake area of the aquifer. The intake areas or recharge areas of confined aquifers are commonly at high-level surface outcrops of the formations. Ground water flow occurs from high-level outcrop areas to low-level outcrop areas, which are areas of natural discharge. It also flows toward points where water levels are lowered artificially by pumping from wells. When the water level in the well stands above the top of the aquifer, the well is described as artesian. A well that yields water by artesian pressure at the ground surface is a flowing artesian well.

Gravity springs occur where water percolating laterally through permeable material overlying an impermeable stratum comes to the surface. They also occur where the land surface intersects the water table. This type of spring is particularly sensitive to seasonal fluctuations in ground water storage and frequently dwindles to a seep or disappears during dry periods. Gravity springs are characteristically low-discharge sources, but when properly developed they make satisfactory individual water supply systems.

Artesian springs discharge from artesian aquifers. They may occur where the confining formation over the artesian aquifer is ruptured by a fault or where the aquifer discharges to a lower topographic area. The flow from these springs depends on the difference in recharge and discharge elevations of the aquifer and on the size of the openings transmitting the water. Artesian springs are usually more dependable than gravity springs, but they are particularly sensitive to the pumping of wells developed in the same aquifer. As a consequence, artesian springs may be dried by pumping.

Springs may be further classified by the nature of the passages through which water issues from the source.

Seepage springs are those in which the water seeps out of sand, gravel, or other material that contains many small interstices. The term as used here includes many large springs as well as small ones. Some of the large springs have extensive seepage areas and are usually marked by the presence of abundant vegetation. The water of small seepage springs may be colored or carry an oily scum because of decomposition of organic matter or the presence of iron. Seepage springs may emerge along the top of an impermeable bed, but they occur more commonly where valleys are cut into the zone of saturation of water-bearing deposits. These springs are generally free from harmful bacteria, but they are susceptible to contamination by surface runoff which collects in valleys or depressions.

Tubular springs issue from relatively large channels, such as the solution channels and caverns of limestone, and soluble rocks and smaller channels that occur in glacial drift. They are sometimes referred to as "bold" springs because the water issues freely from one or more large openings. When the water reaches the channels by percolation through sand or other fine-grained material, it is usually free from contamination. When the channels receive surface water directly or receive the indirect effluent of cesspools, privies, or septic tanks, the water must be regarded as unsafe.

Fissure springs issue along bedding, joint, cleavage, or fault planes. Their distinguishing feature is a break in the rocks along which the water passes. Some of these springs discharge uncontaminated water of deep-source origin. A large number of thermal springs are of this type. Fissure springs, however, may discharge water which is contaminated by surface drainage from strata close to the surface.

DEVELOPMENT BY WELLS

When a well is pumped, the level of the water table in the vicinity of the well will be lowered. (See fig. 2.) This lowering or "drawdown" causes the water table or artesian pressure surface, depending on the type of aquifer, to take the shape of an inverted cone called a cone of depression. This cone, with the well at the apex, is measured in terms of the difference between the static water level and the pumping level. At increasing distances from the well, the drawdown decreases until the slope of the cone merges with the static water table. The distance from the well at which this occurs is called the radius of influence. The radius of influence is not constant but tends to continuously expand with continued pumping. At a given pumping rate, the shape of the cone of depression depends on the characteristics of the water-bearing formation. Shallow and wide cones will form in highly permeable aquifers composed of coarse sands or gravel. Steeper and narrower cones will form in less permeable aquifers. As the pumping rate increases, the drawdown increases and consequently the slope of the cone steepens.

The character of the aquifer — artesian or water table — and the physical characteristics of the formation which will affect the shape of the cone include thickness, lateral extent, and the size and grading of sands or gravels. In a material of low permeability such as fine sand or sandy clay, the drawdown will be greater and the radius of influence less than for the same pumpage from very coarse gravel. (See fig. 2.)

For example, when other conditions are equal for two wells, it may be expected that pumping costs for the same pumping rate will be higher for the well surrounded by material of lower permeability because of the greater drawdown.

When the cones of depression overlap, the local water table will be lowered. (See fig. 2.) This requires additional pumping lifts to obtain water from the interior portion of the group of wells. In addition, a wider distribution of the wells over the ground water basin will reduce the cost of pumping and will allow the development of a larger quantity of water.

Yield of Wells

The amount of water that can be pumped from any well depends on the character of the aquifer and the construction of the well.

Contrary to popular belief, doubling the diameter of a well

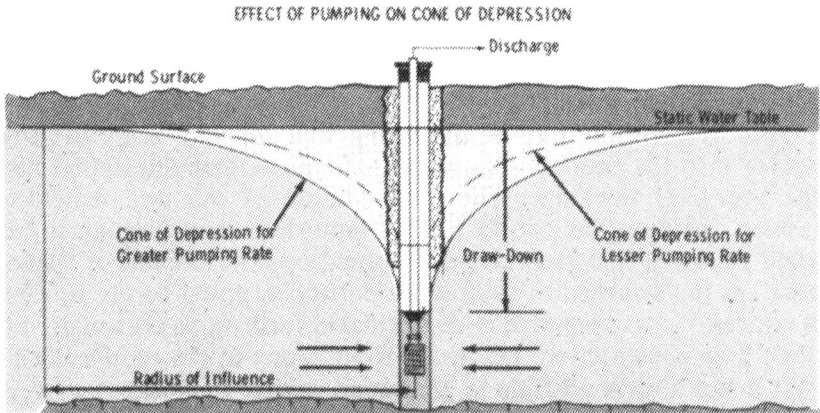

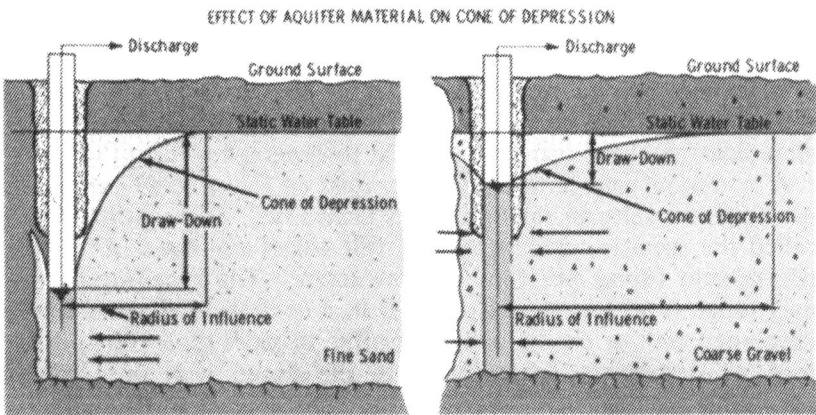

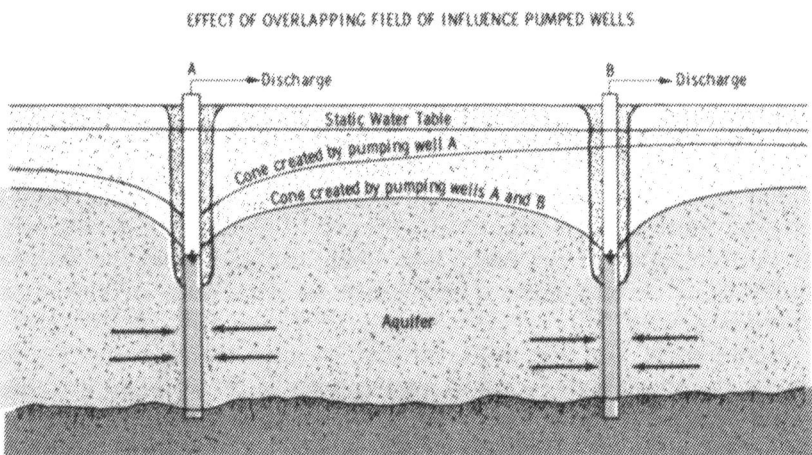

FIGURE 2. Pumping effects on aquifers.

increases its yield only about 10 percent. Or, it could be said that it decreases the drawdown only about 10 percent at the same pumping rate. The casing diameter should be chosen to provide enough room for proper installation of the pump. Individual wells seldom require casings larger than 6 inches. Four-inch wells are common in many areas.

A more effective way of increasing well capacity is by drilling deeper into the aquifer — assuming, of course, that the aquifer has the necessary thickness. The inlet portion of the well structure (screen, perforations, slots) is also important in determining the yield of a well in a sand or gravel formation. The amount of "open area" in the screened or perforated portion exposed to the aquifer is critical. Wells completed in consolidated formations are usually of open hole construction; i.e., there is no casing in the aquifer itself.

It is not always possible to predict accurately the yield of a given well before its completion. Knowledge can be gained, however, from studying the geology of the area and interpreting the results obtained from other wells constructed in the vicinity. This information will be helpful in selecting the location and type of well most likely to be successful. The information can also provide an indication of the quantity or yield to expect.

A common way to describe the yield of a well is to express its discharge capacity in relation to its drawdown. This relationship is called the specific capacity of the well and is expressed in "gallons per minute (gpm) per foot of drawdown." The specific capacity may range from less than 1 gpm per foot of drawdown for a poorly developed well or one in a tight aquifer to more than 100 gpm per foot of drawdown for a properly developed well in a highly permeable aquifer.

Table 3 gives general information on the practicality of penetrating various types of geologic formations by the methods indicated.

Dug wells can be sunk only a few feet below the water table. This seriously limits the drawdown that can be imposed during pumping, which in turn limits the yield of the well. A dug well that taps a highly permeable formation such as gravel may yield 10 to 30 gpm or even more in some situations with only 2 or 3 feet of drawdown. If the formation is primarily fine sand, the yield may be on the order of 2 to 10 gpm. These refer to dug wells of the sizes commonly used.

Bored wells, like dug wells, can also be sunk only a limited depth below the static water level. A penetration of 5 to 10 feet into the water-bearing formation can probably be achieved. If the well is nonartesian, the available drawdown would be 2 or 3 feet less than the depth of water standing in the well. If the well taps an artesian aquifer, however, the static water level will rise to some point above

TABLE 3. — *Suitability of well construction methods to different geological conditions*

| Characteristics | Dug | Bored | Driven | Drilled | | | | Jetted |
| | | | | Percussion | Rotary | | | |
					Hydraulic	Air		
Range of practical depths (general order of magnitude)	0-50 feet	0-100 feet	0-50 feet	0-1,000 feet	0-1,000 feet	0-750 feet		0-100 feet
Diameter	3-20 feet	2-30 inches	1¼-2 inches	4-18 inches	4-24 inches	4-10 inches		2-12 inches
Type of geologic formation:								
Clay	Yes	Yes	Yes	Yes	Yes	No		Yes
Silt	Yes	Yes	Yes	Yes	Yes	No		Yes
Sand	Yes	Yes	Yes	Yes	Yes	No		Yes
Gravel	Yes	Yes	Fine	Yes	Yes	No		¼-inch pea gravel
Cemented gravel	Yes	No	No	Yes	Yes	No		No
Boulders	Yes	Yes, if less than well diameter	No	Yes, when in firm bedding	(Difficult)	No		No
Sandstone . . . . }	Yes, if soft and/or fractured	Yes, if soft and/or fractured }	Thin layers only	Yes	Yes	Yes		No
Limestone . . . . }			No	Yes	Yes	Yes		No
Dense igneous rock	No	No	No	Yes	Yes	Yes		No

¹The ranges of values in this table are based upon general conditions. They may be exceeded for specific areas or conditions.

the top of the aquifer. This rise of the static water level increases the depth of the water. The available drawdown and the yield of the well will therefore be increased. A bored well tapping a highly permeable aquifer and providing several feet of available drawdown may yield 20 gpm or more. If the aquifer has a low permeability or the depth of water in the well is small, the yield may be much lower.

Driven wells can be sunk to as much as 30 feet or more below the static water level. A well at this depth can provide 20 feet or more of drawdown when being pumped. The small diameter of the well, however, limits the type of pump that can be employed, so that the yield under favorable conditions is limited to about 30 gpm. In fine sand or sandy clay formations of limited thickness, the yield may be less than 5 gpm.

Drilled and jetted wells can usually be sunk to such depths that the depth of water standing in the well and consequently the available drawdown will vary from less than 10 to hundreds of feet. In productive formations of considerable thickness, yields of 300 gpm and more are readily attained. Drilled wells can be constructed in all instances where driven wells are used and in many areas where dug and bored wells are constructed. The larger diameter of a drilled well as opposed to that of a driven well permits use of larger pumping equipment that can develop the full capacity of the aquifer. As has already been pointed out, the capacity or yield of a well varies greatly, depending upon the permeability and thickness of the formation, the construction of the well, and the available drawdown.

Preparation of Ground Surface at Well Site

A properly constructed well should exclude surface water from a ground water source to the same degree as does the undisturbed overlying geologic formation. The top of the well must be constructed so that no foreign matter or surface water can enter. The well site should be properly drained and adequately protected against erosion, flooding, and damage or contamination from animals. Surface drainage should be diverted away from the well.

CONSTRUCTION OF WELLS

Dug Wells

The dug well, constructed by hand, is usually shallow. It is more difficult to protect from contamination, although if finished properly it may provide a satisfactory supply. Because of advantages offered by other types of wells, consideration should first be given to one of those described in this section.

Dug wells are usually excavated with pick and shovel. The excavated material can be lifted to the surface by a bucket attached to a windlass or hoist. A power-operated clam shell or orange peel bucket may be used in holes greater than 3 feet in diameter where the material is principally gravel or sand. In dense clays or cemented materials, pneumatic hand tools are effective means of excavation.

To prevent the native material from caving, one must place a crib or lining in the excavation and move it downward as the pit is deepened. The space between the lining and the undisturbed embankment should be backfilled with clean material. In the region of water-bearing formations, the backfilled material should be sand or gravel. Cement grout should be placed to a depth of 10 feet below the ground surface to prevent entrance of surface water along the well lining. (See fig. 3.)

Dug wells may be lined with brick, stone, or concrete, depending on the availability of materials and the cost of labor. Precast concrete pipe, available in a wide range of sizes, provides an excellent lining. This lining can be used as a crib as the pit is deepened. When the lining is to be used as a crib, concrete pipe with tongue-and-groove joints and smooth exterior surface is preferred. (See fig. 3.)

Bell and spigot pipe may be used for a lining where it can be placed inside the crib or in an unsupported pit. This type of pipe requires careful backfilling to guarantee a tight well near the surface. The prime factor with regard to preventing contaminated water from entering a dug well is the sealing of the well lining and otherwise excluding draining-in of surface water at and near the well.

Most dug wells do not penetrate much below the water table because of the difficulties in manual excavation and the positioning of cribs and linings. The depth of excavation can be increased by the use of pumps to lower the water level during construction. Because of their shallow penetration into the zone of saturation, many dug wells fail in times of drought when the water level recedes or when large quantities of water are pumped from the wells.

Bored Wells

Bored wells are commonly constructed with earth augers turned either by hand or by power equipment. Such wells are usually regarded as practical at depths of less than 100 feet when the water requirement is low and the material overlying the water-bearing formation has noncaving properties and contains few large boulders. In suitable material, holes from 2 to 30 inches in diameter can be bored to about 100 feet without caving.

In general, bored wells have the same characteristics as dug wells, but they may be extended deeper into the water-bearing formation.

Bored wells may be cased with vitrified tile, concrete pipe,

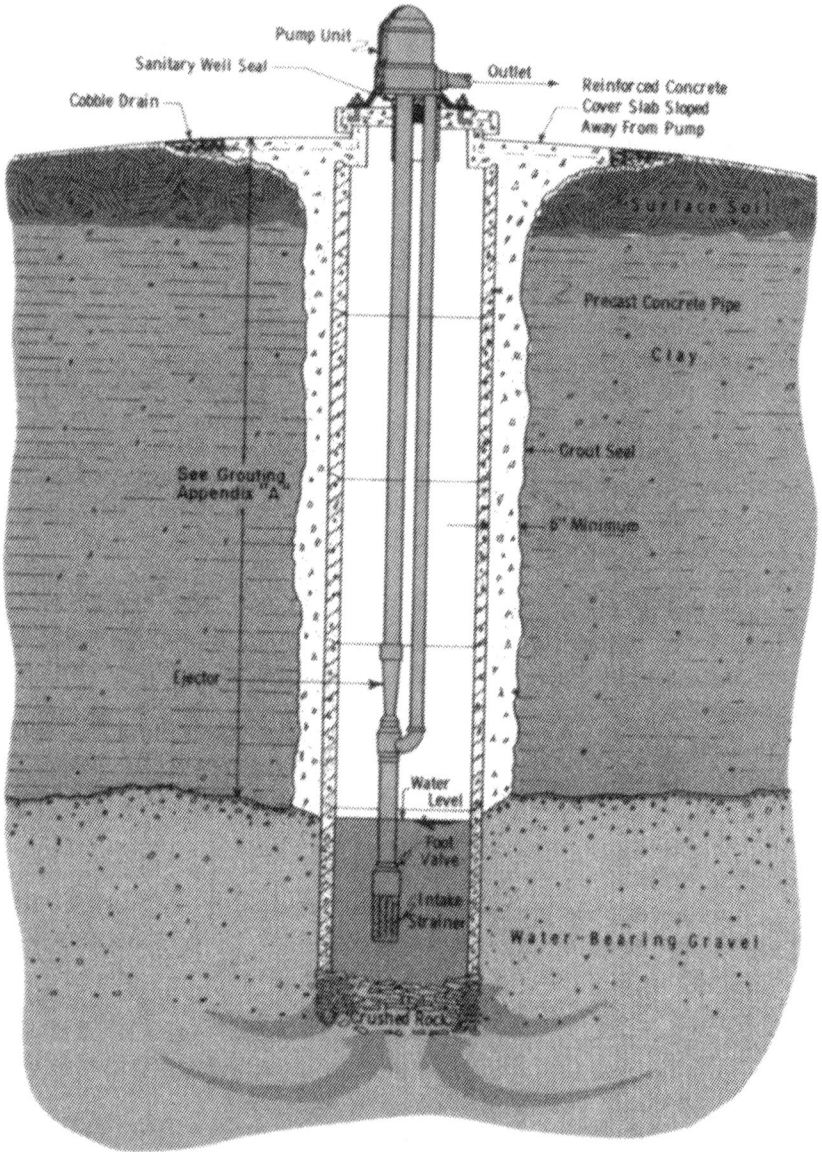

FIGURE 3. Dug well with two-pipe jet pump installation.

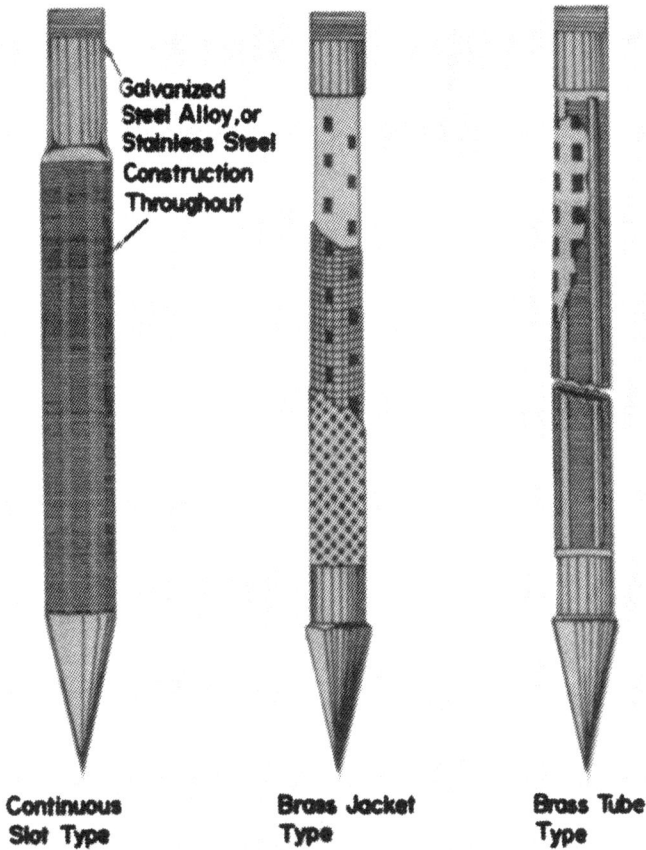

Galvanized
Steel Alloy, or
Stainless Steel
Construction
Throughout

Continuous
Slot Type

Brass Jacket
Type

Brass Tube
Type

FIGURE 4. Different kinds of drive-well points.

standard wrought iron, steel casing, or other suitable material capable of sustaining imposed loads. The well may be completed by installing well screens or perforated casing in the water-bearing sand and gravel. Proper protection from surface drainage should be provided by sealing the casing with cement grout to the depth necessary to protect the well from contamination. (See p. 48 and app. A.)

Driven Wells

The simplest and least expensive of all well types is the driven well. It is constructed by driving into the ground a drive-well point which is fitted to the end of a series of pipe sections. (See figs. 4-5.) The drive point is of forged or cast steel. Drive points are usually 1¼ or 2 inches in diameter. The well is driven with the aid of a maul, or a special drive weight. (See fig. 5.) For deeper wells, the well points are sometimes driven into water-bearing strata from the bottom of a bored or dug well. (See fig. 6.) The yield of driven

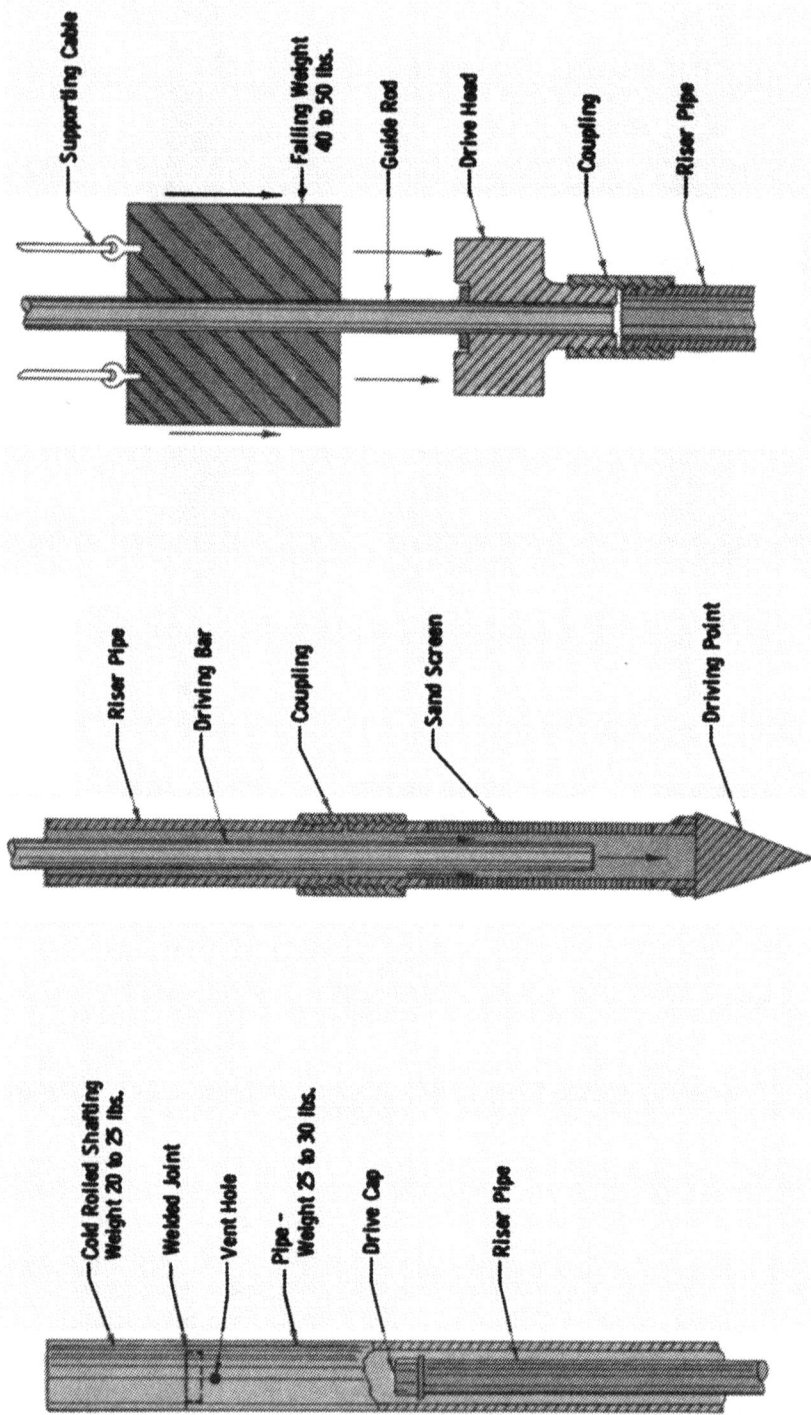

Supporting Cable

Falling Weight
40 to 50 lbs.

Guide Rod

Drive Head

Coupling

Riser Pipe

Riser Pipe

Driving Bar

Coupling

Sand Screen

Driving Point

Cold Rolled Shafting
Weight 20 to 25 lbs.

Welded Joint

Vent Hole

Pipe -
Weight 25 to 30 lbs.

Drive Cap

Riser Pipe

FIGURE 5. Well-point driving methods.

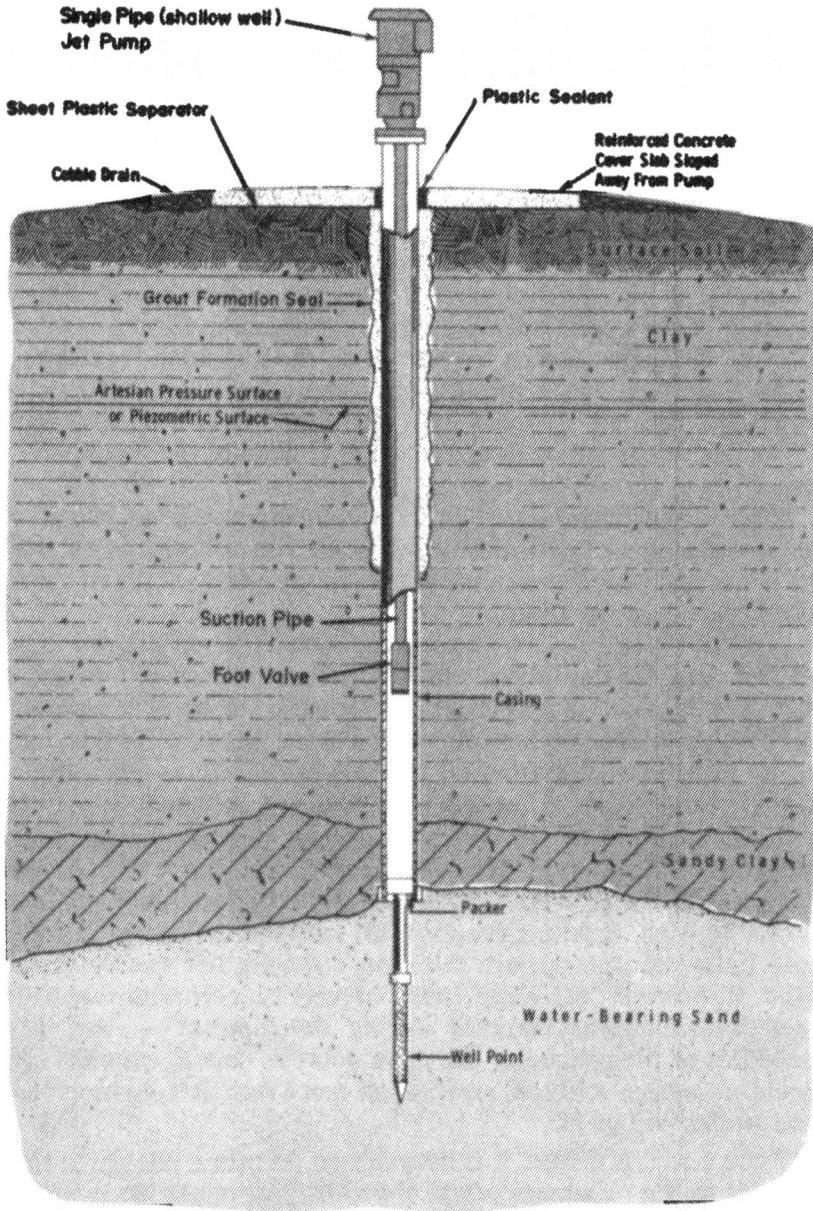

FIGURE 6. Hand-bored well with driven well point and "shallow well" jet pump.

wells is generally small to moderate. Where they can be driven an appreciable depth below the water table, they are no more likely than bored wells to be seriously affected by water-table fluctuations. The most suitable locations for driven wells are areas containing alluvial deposits of high permeability. The presence of coarse gravels, cobbles, or boulders interferes with sinking the well point and may damage the wire mesh jacket.

Well-drive points can be obtained in a variety of designs and materials. (See fig. 4.) In general, the serviceability and efficiency of each is related to its basic design. The continuous-slot, wire-wound type is more resistant to corrosion and can usually be treated with chemicals to correct problems of incrustation. It is more efficient because of its greater open area, and is easier to develop (see p. 44) because its design permits easy access to the formation for cleanup.

Another type has a metal gauze wrapped around a perforated steel pipe base and covered by a perforated jacket; if it contains dissimilar metals, electrolytic corrosion is likely to shorten its life — especially in corrosive waters.

Wherever maximum capacity is required, well-drive points of good design are a worthwhile investment. The manufacturer should be consulted for his recommendation of the metal alloy best suited to the particular situation.

Good drive-well points are available with different size openings, or slot sizes, for use in sands of different grain sizes. If too large a slot size is used, it may never be possible to develop the well properly, and the well is likely to be a "sand pumper," or gradually to fill in with sand, cutting off the flow of water from the aquifer. On the other hand, if the slot size chosen is too small, it may be difficult to improve the well capacity by development, and the yield may be too low. When the nature of aquifer sand is not known beforehand, a medium-sized slot — 0.015 inch or 0.020 inch — can be tried. If during development sand and sediments continue indefinitely to pass through the slots, a smaller slot size should be used. If, however, the water cleans up very quickly with very little sand and sediment removed during development — less than one-third of the volume of the drive point — then a larger slot size could have been selected, resulting in more complete development and greater well yield.

When a well is driven, it is desirable to prepare a pilot hole that extends to the maximum practical depth. This can be done with a hand auger slightly larger than the well point. After the pilot hole has been prepared, the assembled point and pipe are lowered into the hole. Depending on the resistance afforded by the formation, driving is accomplished in several ways. The pipe is driven by directly striking the drive cap, which is snugly threaded to the top of the protruding section of the pipe. A maul, a sledge, or a special driver may be used to hand-drive the pipe. The special driver may consist of a weight and sleeve arrangement which slides over the drive cap as the weight is lifted and dropped in the driving process. (See fig. 5.)

Jetted Wells

A rapid and efficient method of sinking well points is that of jetting or washing-in. This method requires a source of water and a pressure pump. Water forced under pressure down the riser pipe issues from a special washing point. The well point and pipe are then lowered as material is loosened by the jetting.

The riser pipe of a jetted well is often used as the suction pipe for the pump. In such instances, surface water may be drawn into the well if the pipe develops holes by corrosion. An outside protective casing may be installed to the depth necessary to provide protection against the possible entry of contaminated surface water. The annular space between the casings should then be filled with cement grout. The protective casing is best installed in an auger hole and the drive point then driven inside it.

Drilled Wells

Construction of a drilled well (see fig. 7) is ordinarily accomplished by one of two techniques – percussion or rotary hydraulic drilling. The selection of the method depends primarily on the geology of the site and the availability of equipment.

Percussion (Cable-Tool) Method. Drilling by the cable-tool or percussion method is accomplished by raising and dropping a heavy drill bit and stem. The impact of the bit crushes and dislodges pieces of the formation. The reciprocating motion of the drill tools mixes the drill cuttings with water into a slurry at the bottom of the hole. This is periodically brought to the surface with a bailer, a 10- to 20-foot-long pipe equipped with a valve at the lower end.

Caving is prevented as drilling progresses by driving or sinking into the ground a casing slightly larger in diameter than the bit. When wells are drilled in hard rock, casing is usually necessary only through the overburden of unconsolidated material. A casing may be necessary in hard rock formations to prevent caving of beds of softer material.

When good drilling practices are followed, water-bearing beds are readily detected in cable-tool holes, because the slurry does not tend to seal off the water-bearing formation. A rise or fall in the water level in the hole during drilling, or more rapid recovery of the water level during bailing, indicates that a permeable bed has been entered. Crevices or soft streaks in hard formations are often water bearing. Sand, gravel, limestone, and sandstone are generally permeable and yield the most water.

Hydraulic Rotary Drilling Method. The hydraulic rotary drilling

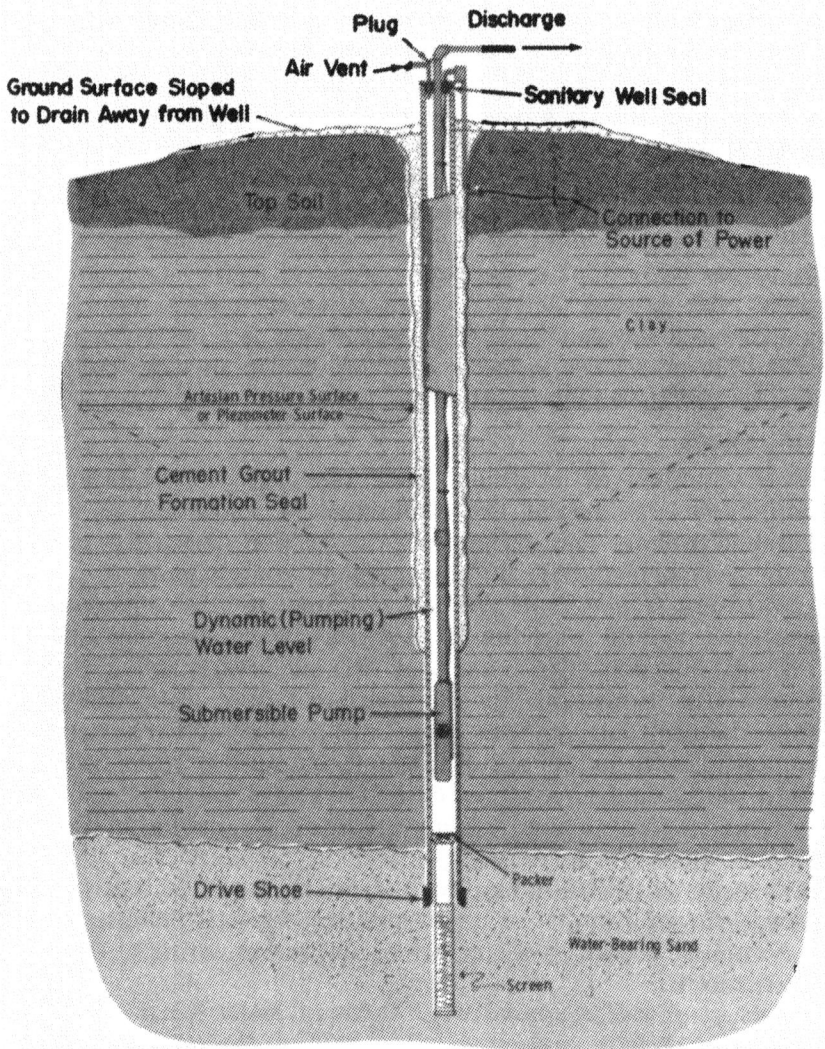

FIGURE 7. Drilled well with submersible pump.

method may be used in most formations. The essential parts of the drilling assembly include a derrick and hoist, a revolving table through which the drill pipe passes, a series of drill-pipe sections, a cutting bit at the lower end of the drill pipe, a pump for circulation of drilling fluid, and a power source to drive the drill.

In the drilling operation, the bit breaks up the material as it rotates and advances. The drilling fluid (called mud) pumped down the drill pipe picks up the drill cuttings and carries them up the annular space between the rotating pipe and the wall of the hole. The mixture of mud and cuttings is discharged to a settling pit where the cuttings drop to the bottom and mud is recirculated to the drill pipe.

When the hole is completed, the drill pipe is withdrawn and the casing placed. The drilling mud is usually left in place and pumped out after the casing and screen are positioned. The annular space between the hole wall and the casing is generally filled with cement grout in non-water-bearing sections, but may be enlarged and filled with gravel at the level of water-bearing strata.

When little is known concerning the geology of the area, the search for water-bearing formations must be done carefully and deliberately so that all possible formations are located and tested. Water-bearing formations may be difficult to recognize by the rotary method or may be plugged by the pressure of the mud.

Air Rotary Drilling Method. The air rotary method is similar to the rotary hydraulic method in that the same type of drilling machine and tools may be used. The principal difference is that air is the fluid used rather than mud or water. In place of the conventional mud pump to circulate the fluids, air compressors are used. Many drillers equip the rig with a mud pump to increase the versatility of the equipment.

The air rotary method is well adapted to rapid penetration of consolidated formations, and is especially popular in regions where limestone is the principal source of water. It is not generally suited to unconsolidated formations where careful sampling of rock materials is required for well-screen installation. Small quantities of water can be detected readily during drilling, and the yield estimated. Larger sources of water may impede progress.

The air rotary method requires that air be supplied at pressures from 100 to 250 pounds per square inch. To effect removal of the cuttings, ascending velocities of at least 3,000 feet per minute are necessary.

Penetration rates of 20 to 30 feet per hour in very hard rock are common with air rotary methods.

Conventional mud drilling is sometimes used to drill through caving formations that overlie bedrock. Casing may have to be installed through the overburden before continuing with the air rotary method.

Down-the-Hole Air Hammer The down-hole pneumatic hammer combines the percussion effect of cable-tool drilling and the rotary movement of rotary drilling. The tool bit is equipped with tungsten-carbide inserts at the cutting surfaces. Tungsten-carbide is very resistant to abrasion.

Water Well Casing and Pipe

There are several kinds of steel pipe that are suitable for casing drilled wells. The following are the most commonly used:

Standard pipe
Line pipe
Drive pipe
Reamed and drifted (R&D) pipe
Water well casing

There are certain differences in sizes, in wall thicknesses, in types of threaded connections available, and in methods of manufacture. The important thing for the owner to know about well casing is that it meet certain generally accepted specifications for quality of the steel and thickness of the wall. Both are important because they determine resistance to corrosion, and consequently the useful life of the well. Strength of the casing may also be important in determining whether certain well construction procedures may be successfully carried out – particularly in cable-tool drilling where hard driving of the casing is sometimes required.

The most commonly accepted specifications for water well casing are those prepared by:

American Society for Testing and Materials (ASTM)
American Petroleum Institute (API)
American Iron and Steel Institute (AISI)
Federal Government

Each source lists several specifications that might be used, but those most likely to be called for are ASTM A-120 and A-53, API 5-L, AISI Standard for R&D pipe, and Federal specification WW-P-406B.

Table 4 lists "standard weight" wall thicknesses for standard pipe and line pipe through the sizes ordinarily used in well construction. Thinner pipe should not be used. If conditions in the area are known to be highly corrosive, the "extra strong" and heavier weights should be used.

Setting Screens or Slotted Casings in Wells

Screens or slotted casings are installed in wells to permit sand-free water to flow into the well and to provide support for unstable formations to prevent caving. The size of the slot for the screen or perforated pipe should be based on a sieve analysis of carefully selected samples of the water-bearing formation that is to be developed. The analysis is usually made by the screen manufacturer.

42

TABLE 4. – *Steel pipe and casing, standard and standard line pipe*

Nominal size (in.)	Diameters (in.)		Wall thickness (in.)	Approximate weight (lb./ft.)	
	Outside	Inside		Plain ends	Threaded and coupled
1¼	1.660	1.380	.140	2.27	2.30
1½	1 900	1 610	.145	2 72	2.75
2	2 375	2.067	.154	3.65	3.75
3	3 500	3 068	.216	7.58	7.70
4	4 500	4.026	.237	10.79	11.00
5	5.563	5 047	.258	14 62	15 00
6	6 625	6 065	.280	18.97	19.45
8	8 625	8.071	.277	24.70	25.55
8	8.625	7.981	.322	28 55	29.35
10	10 750	10.192	.279	31 20	32.75
10	10.750	10.136	.307	34.24	35.75
10	10.750	10.020	365	40.48	41.85
12	12.750	12.090	.330	43.77	45.45
12	12.750	12 000	.375	49.56	51 10

If the slot size is too large, the well may yield sand when pumped. If the slots are too small, they may become plugged with fine material and the well yield will be reduced. In a drilled well, the screens are normally placed after the casing has been installed; however, in a driven well, the screen is a part of the drive assembly and is sunk to its final position as the well is driven.

The relationship between the open area of the screen and the velocity of water through the openings should be considered if maximum hydraulic efficiency is desired. Loss of energy through friction is kept to a minimum by holding velocities to 0.1 foot per second or less. Since the slot size is determined by the grain size distribution in the aquifer sand, the required open area must be obtained by varying the diameter – or, if aquifer thickness permits, the length – of the screen. Manufacturers of well screens provide tables of capacities and other information to facilitate selection of the most economical screen dimensions.

Methods of screen installation in drilled wells include (1) the pullback method, (2) the open-hole method, and (3) the baildown method. The pullback method of installation is one in which the casing is drawn back to expose a well screen placed inside the casing at the bottom of the well. In the open-hole installation the screen attached to the casing is inserted in the uncased bottom part of the hole when the aquifer portion of the hole remains open. When the baildown method is employed, the screen is placed at the bottom of the cased hole and advanced into the water-bearing formation by bailing the sand out from below the screen.

The *pullback* method is suited to bored or drilled wells, as long as the casing can be moved, while the *open-hole* method is used in most instances with rotary drilling. The *baildown* method may be used in wells drilled by any method where water-bearing formations consist of sand. It is not well adapted to gravel formations.

A screen is seldom required in wells tapping bedrock or tightly cemented sediments such as sandstone or limestone.

A fourth method, adaptable primarily in rotary drilled holes, is the washdown method. This procedure entails the circulation of water, by use of the mud pump, through a special self-closing bottom upward around the screen and through the annular space between the washpipe and the permanent casing to the surface. As material is washed by jet action from below it, the well screen settles to its desired position.

If the screen is placed after positioning of the casing, it must be firmly sealed to the casing. This is generally done by swaging out a lead packer attached to the top of the screen. When the pullback method of installation is employed, a closed bail bottom usually provides the bottom closure; a self-closing bottom serves this purpose when the washdown method is used. A special plug is placed in the bottom when the baildown method is employed. A quantity of lead wool or a small bag of dry cement may also be tamped into the bottom of the screen to seal it.

Development of Wells

Before a well is put into use, it is necessary to completely remove silt and fine sand from the formation adjacent to the well screen by one of several processes known as "development." The development procedure unplugs the formation and produces a natural filter of coarser and more uniform particles of high permeability surrounding the well screen. After the development is completed, there will be a well-graded, stabilized layer of coarse material which will entirely surround the well screen and facilitate the flow of water in the formation into the well.

The simplest method of well development is that of surging. In this process the silt and sand grains are agitated by a series of rapid reversals in the direction of flow of water and are drawn toward the screen through larger pore openings. A well may be surged by moving a plunger up and down in it. This action moves the water alternately into and out of the formation. When water containing fine granular material moves into the well, the particles tend to settle to the bottom of the screen. They can be removed subsequently by pumping or bailing.

One of the most effective methods of development is the high-velocity hydraulic-jetting method. Water under pressure ejected from orifices passes through the slot openings, violently agitating the aquifer material. Sand grains finer than the slot size move through the screen and either settle to the bottom of the well (from which they are subsequently removed by bailing) or are washed out at the top (if the well overflows). Conventional centrifugal or piston pumps may be used; or the mud pump of the rotary hydraulic drill easily accomplishes this. Pressures of at least 100 psi should be used, with pressure greater than 150 psi preferred. In addition to the intensity of development that may be applied by this method, it has the advantage of permitting selective concentration of development on those portions of the screen most in need. High-velocity jetting is of most benefit in screens of continuous horizontal slot design. It has also proven effective in washing out drilling mud and cuttings from crevices in hard-rock wells. It is less useful in slotted or perforated pipe.

Other methods of development are interrupted pumping and, sometimes in consolidated material, explosives when used only by experts. The method of development must be suited to the aquifer and the type of well construction. Proper development is necessary in many formations and under many conditions for the completion of a successful well. Its importance should not be overlooked.

Testing Well for Yield and Drawdown

In order that the most suitable pumping equipment can be selected, a pumping test should be made after the well has been developed to determine its yield and drawdown. The pumping test for yield and drawdown should include the determination of the volume of water pumped per minute or hour, the depth to the pumping level as determined over a period of time at one or more constant rates of pumpage, the recovery of the water level after pumping is stopped, and the length of time the well is pumped at each rate during the test procedure. When the completed well is tested for yield and drawdown, it is essential that it be done accurately by the use of approved measuring devices and accepted methods. Additional information regarding the testing of wells for drawdown or yield may be obtained from the U.S. Geological Survey, the State or local health department, and the manufacturers of well screens or pumping equipment.

Water table wells (see pp. 26, 28) are more affected than artesian wells by seasonal fluctuations in ground water levels. When testing a water table well for yield and drawdown, it is desirable — though frequently not practical — to test it near the end of the dry season. When this cannot be done, it is important to determine as nearly as possible, from other wells tapping the same formations, the additional seasonal decline in water levels that can be expected. This additional decline should then be added to the drawdown determined by the pumping test, in order to arrive at the ultimate pumping water level. Seasonal declines of several feet in water table wells are not unusual, and these can seriously reduce the capacity of such wells in the dry season.

Individual wells should be test pumped at a constant pumping rate that is not less than that planned for the final pump installation. The well should be pumped at this rate for not less than 4 hours, and the maximum drawdown recorded. Measurements of the water levels during recovery can then be made. Failure to recover completely to the original static water level within 24 hours should be reason to question the dependability of the water-bearing formation.

Well Failure

Over a period of time, wells may fail to produce for any of these main causes:

1. Failure or wear of the pump.
2. Declining water levels.
3. Plugged or corroded screens.
4. Accumulation of sand or sediments in the well.

Proper analysis of the cause necessitates measuring the water level before, during, and after pumping. To facilitate measuring the water level, one should provide for the entrance of a tape or an electrical measuring device into the well in the annular space between the well casing and the pump column (figs. 7-8).

An "air line" with a water-depth indicating gage, available from pump suppliers, may also be used. On some larger wells, the air line

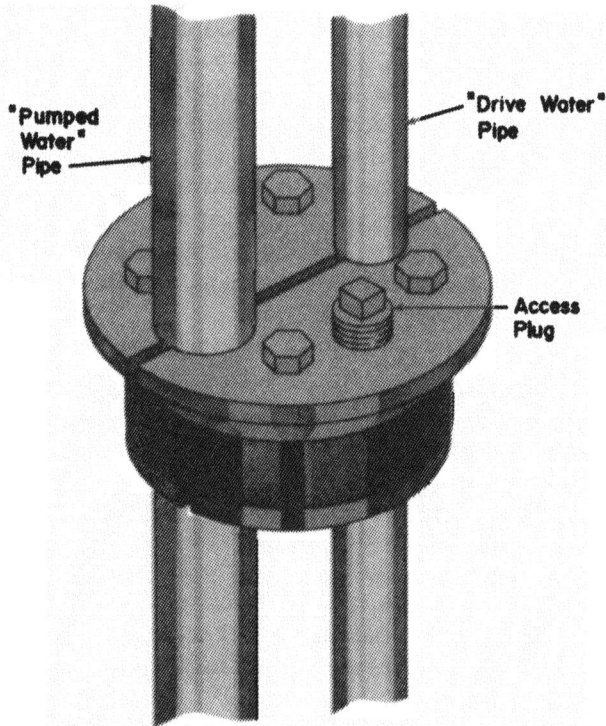

FIGURE 8. Well seal for jet pump installation.

and gage are left installed so that periodic readings can be taken and a record of well and pump performance kept. While not as accurate as the tape or electrode method, this installation is popular for use in a well that is being pumped because it is unaffected by water that may be falling from above.

Unless the well is the pitless adapter or pitless unit type (p. 109), access for water-level measurements can be obtained through a threaded hole in the sanitary well seal (figs. 8-9). This is true for submersible and jet pump installations, as well as for some others. If it is not possible to gain access through the top of the well, access may be provided by means of a pipe welded to the side of the casing. (See discussion under "Installation of Pumping Equipment," p. 104.)

If the well is completed as a pitless adapter installation (p. 109),

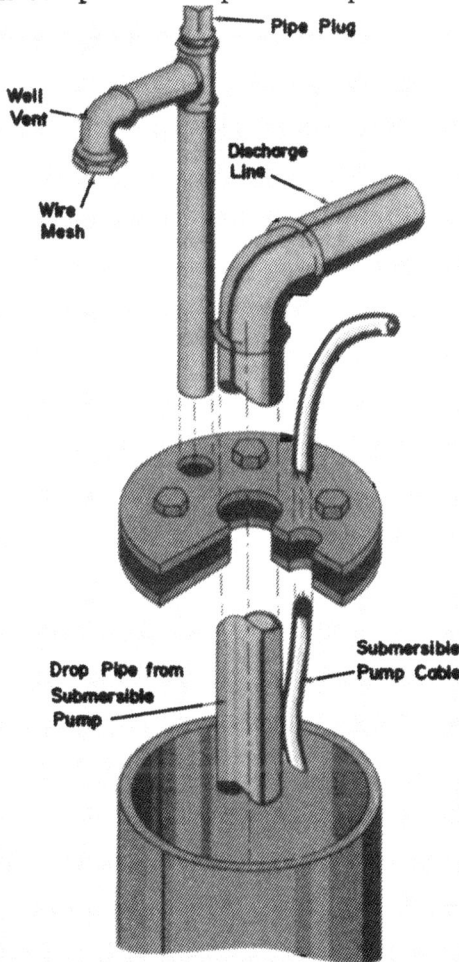

FIGURE 9. Well seal for submersible pump installation.

it is usually possible to slide the measuring device past the adapter assembly inside the casing and on to the water below. If it is a pitless unit, particularly the "spool" type (see fig. 22, p. 112), it probably will not be possible to reach the water level. In the latter case, the well can only be tested by removing the spool and pump and reinstalling the pump, or another one, without the spool.

Any work performed within the well — including insertion of a measuring line — is likely to contaminate the water with coliform bacteria and other organisms. The well should be disinfected (see p. 50) before returning it to service. All access holes should be tightly plugged or covered following the work.

Sanitary Construction of Wells

The penetration of a water-bearing formation by a well provides a direct route for possible contamination of the ground water. Although there are different types of wells and well construction, there are basic sanitary aspects that must be considered and followed.

1. The annular space outside the casing should be filled with a watertight cement grout or puddled clay from a point just below the frost line or deepest level of excavation near the well (see "Pitless Units and Adapters," p.109), to as deep as necessary to prevent entry of contaminated water. See appendix A for grouting recommendations.

2. For artesian aquifers, the casing should be sealed into the overlying impermeable formations so as to retain the artesian pressure.

3. When a water-bearing formation containing water of poor quality is penetrated, the formation should be sealed off to prevent the infiltration of water into the well and aquifer.

4. A sanitary well seal with an approved vent should be installed at the top of the well casing to prevent the entrance of contaminated water or other objectionable material.

For large-diameter wells such as dug wells, it would be difficult to provide a sanitary well seal; consequently, a reinforced concrete slab, overlapping the casing and sealed to it with a flexible sealant or rubber gasket, should be installed. The annular space outside the casing should first be filled with suitable grouting or sealing materials — cement, clay, or fine sand.

Well Covers and Seals

Every well should be provided with an overlapping, tight-fitting cover at the top of the casing or pipe sleeve to prevent contaminated water or other material from entering the well.

The sanitary well seal in a well exposed to possible flooding should be either watertight or elevated at least 2 feet above the highest known flood level. When it is expected that a well seal may become flooded, it should be watertight and equipped with a vent line whose opening to the atmosphere is at least 2 feet above the highest known flood level.

The seal in a well *not* exposed to possible flooding should be either watertight (with an approved vent line) or self-draining with an overlapping and downward flange. If the seal is of the self-draining (nonwatertight) type, all openings in the cover should be either watertight or flanged upward and provided with overlapping, downward flanged covers.

Some pump and power units have closed bases that effectively seal the upper terminal of the well casing. When the unit is the open type, or when it is located at the side (some jet- and suction-pump-type installations), it is especially important that a sanitary well seal be used. There are several acceptable designs consisting of an expandable neoprene gasket compressed between two steel plates (see figs. 8-9). They are easily installed and removed for well servicing. Pump and water well suppliers normally stock sanitary well seals.

If the pump is not installed immediately after well drilling and placement of the casing, the top of the casing should be closed with a metal cap screwed or tack-welded into place, or covered with a sanitary well seal.

A well slab alone is not an effective sanitary defense, since it can be undermined by burrowing animals and insects, cracked from settlement or frost heave, or broken by vehicles and vibrating machinery. The cement grout formation seal is far more effective. (See p. 48.) It is recognized, however, that there are situations that call for a concrete slab or floor around the well casing to facilitate cleaning and improve appearance. When such a floor is necessary, it should be placed only after the formation seal and the pitless installation (see p. 109) have been inspected.

Well covers and pump platforms should be elevated above the adjacent finished ground level. Pumproom floors should be constructed of reinforced, watertight concrete, and carefully leveled or sloped away from the well so that surface and waste water cannot stand near the well. The minimum thickness of such a slab or floor should be 4 inches. Concrete slabs or floors should be poured separately from the cement formation seal and — when the threat of freezing exists — insulated from it and the well casing by a plastic or mastic coating or sleeve to prevent bonding of the concrete to either.

All water wells should be readily accessible at the top for inspection, servicing, and testing. This requires that any structure over the well be easily removable to provide full, unobstructed access for well-servicing equipment. The so-called "buried seal," with the well cover buried under several feet of earth, is unacceptable because (1) it discourages periodic inspection and preventive maintenance, (2) it makes severe contamination during pump servicing and well repair more likely, (3) any well servicing is more expensive, and (4) excavation to expose the top of the well increases the risk of damage to the well, the cover, the vent and the electrical connections.

Disinfection of Wells

All newly constructed wells should be disinfected to neutralize contamination from equipment, material, or surface drainage introduced during construction. Every well should be disinfected promptly after construction or repair.

An effective and economical method of disinfecting wells and appurtenances is that of using calcium hypochlorite containing approximately 70-percent available chlorine. This chemical can be purchased in granular or tablet form at hardware stores, swimming pool equipment supply outlets, or chemical supply houses.

When used in the disinfection of wells, calcium hypochlorite should be added in sufficient amounts to provide a dosage of approximately 100 mg/ℓ of available chlorine in the well water. This concentration is roughly equivalent to a mixture of 2 ounces of dry chemical per 100 gallons of water to be disinfected. Practical disinfection requires the use of a stock solution. The stock solution may be prepared by mixing 2 ounces of high-test hypochlorite with 2 quarts of water. Mixing is facilitated if a small amount of the water is first added to the granular calcium hypochlorite and stirred to a smooth watery paste free of lumps. It should then be mixed with the remaining quantity of water. The stock solution should be stirred thoroughly for 10 to 15 minutes prior to allowing the inert ingredients to settle. The clearer liquid containing the chlorine should be used and the inert material discarded. Each 2 quarts of stock solution will provide a concentration of approximately 100 mg/ℓ when added to 100 gallons of water. The solution should be prepared in a thoroughly clean utensil; the use of metal containers should be avoided, if possible, as they are corroded by strong chlorine solutions. Crockery, glass, or rubber-lined containers are recommended.

Where small quantities of disinfectant are required and a scale is not available, the material can be measured with a spoon. A heaping tablespoonful of granular calcium hypochlorite weighs approximately ½ ounce.

When calcium hypochlorite is not available, other sources of available chlorine, such as sodium hypochlorite (12-15 percent of volume), can be used. Sodium hypochlorite, which is also commonly available as liquid household bleach with 5.25 percent available chlorine, can be diluted with one part of water to produce the stock solution. Two quarts of this solution can be used for disinfecting 100 gallons of water.

Stock solutions of chlorine in any form will deteriorate rapidly unless properly stored. Dark glass or plastic bottles with airtight caps are recommended. Bottles containing solution should be kept in a cool place and protected from direct sunlight. If proper storage facilities are not available, the solution should always be prepared fresh immediately before use. Commercially available household bleach solutions, because of their convenience and usual reliability as to concentration or strength, are preferred stock solutions for disinfecting individual water supplies.

Table 5 shows quantities of disinfectants to be used in treating wells of different diameters and water depths. For sizes or depths not shown, the next larger figure should be used.

Dug Wells

1. After the casing or lining has been completed, follow the procedure outlined below before placing the cover platform over the well.

    a. Remove all equipment and materials, including tools, forms, platforms, etc., that will not form a permanent part of the completed structure.

    b. Using a stiff broom or brush, wash the interior wall of the casing or lining with a strong solution (100 mg/ℓ of chlorine) to insure thorough cleaning.

2. Place the cover over the well and pour the required amount of chlorine solution into the well through the manhole or pipesleeve opening just before inserting the pump cylinder and drop-pipe assembly. The chlorine solution should be distributed over as much of the surface of the water as possible to obtain proper diffusion of the chemical through the water. Diffusion of the chemical with the well water may be facilitated by running the solution into the well through the water hose or pipeline as the line is being alternately raised and lowered. This method should be followed whenever possible.

3. Wash the exterior surface of the pump cylinder and drop pipe with the chlorine solution as the assembly is being lowered into the well.

4. After the pump has been set in position, pump water from the well until a strong odor of chlorine is noted.

5. Allow the chlorine solution to remain in the well for not less than 24 hours.

6. After 24 hours or more have elapsed, flush the well to remove all traces of chlorine.

TABLE 5. — Quantities[a] of calcium hypochlorite, 70 percent (rows A) and liquid household bleach, 5.25 percent (rows B) required for water well disinfection

Depth of water in well (ft.)		Well diameter (in.)															
		2	3	4	5	6	8	10	12	16	20	24	28	32	36	42	48
5	A	1T	1T	1T	1T	1T	1T	2T	3T	5T	6T	3 oz.	4 oz.	5 oz.	7 oz.	9 oz.	12 oz.
	B	1C	1C	1C	1C	1C	1C	1C	1C	2C	4C	1Q	2Q	3Q	3Q	4Q	5Q
10	A	1T	1T	1T	1T	1T	2T	3T	5T	8T	4 oz.	6 oz.	8 oz.	10 oz.	13 oz.	1¼ lb.	1½ lb.
	B	1C	1C	1C	1C	1C	1C	2C	2C	1Q	2Q	3Q	4Q	4Q	6Q	8Q	2½G
15	A	1T	1T	1T	1T	2T	3T	5T	8T	4 oz.	6 oz.	9 oz.	12 oz.	1 lb.	1½ lb.	1¾ lb.	2¾ lb.
	B	1C	1C	1C	1C	1C	2C	3C	4C	2Q	2½Q	4Q	5Q	6Q	3G	3G	3G
20	A	1T	1T	1T	2T	3T	4T	6T	3 oz.	5 oz.	8 oz.						
	B	1C	1C	1C	1C	1C	2C	4C	1Q	2½Q	3½Q				3G	3G	4G
30	A	1T	1T	2T	3T	4T	6T	3 oz.	4 oz.	8 oz.	12 oz.						
	B	1C	1C	1C	1C	2C	4C	1½Q	2Q	4Q	5Q						
40	A	1T	1T	2T	4T	6T	8T	4 oz.	6 oz.	10 oz.	1 lb.						
	B	1C	1C	1C	1C	2C	1Q	2Q	2½Q	4½Q	7Q						
60	A	1T	2T	3T	5T	8T	4 oz.	6 oz.	9 oz.								
	B	1C	1C	2C	3C	4C	2Q	3Q	4Q								
80	A	1T	3T	4T	7T	9T	5 oz.	8 oz.	12 oz.								
	B	1C	1C	2C	4C	1Q	2Q	3½Q	5Q								
100	A	2T	3T	5T	8T	4 oz.	7 oz.	10 oz.	1 lb.								
	B	1C	2C	3C	1Q	1½Q	2½Q	4Q	6Q								
150	A	3T	5T	8T	4 oz.	6 oz.	10 oz.	1 lb.	1½ lb.								
	B	2C	2C	4C	2Q	2½Q	4Q	6Q	2½G								

[a]Quantities are indicated as: T = tablespoons; oz. = ounces (by weight); C = cups; lb. = pounds; Q = quarts; G = gallons.

NOTE: Figures corresponding to rows A are amounts of solid calcium hypochlorite required; those corresponding to rows B are amounts of liquid household bleach. For cases lying in green-shaded area, add 5 gallons of chlorinated water; add 10 gallons of chlorinated water. For those in the blue-shaded area, to force solution into formation. For those in the blue-shaded area, as final step, to force solution into formation. (See "Disinfection of Wells," pp. 50 ff.)

Drilled, Driven, and Bored Wells

1. When the well is being tested for yield, the testpump should be operated until the well water is as clear and as free from turbidity as possible.

2. After the testing equipment has been removed, slowly pour the required amount of chlorine solution into the well just before installing the permanent pumping equipment. Diffusion of the solution with the well water may be facilitated as previously described in item 2, "Dug Wells."

3. Add 5 or 10 gallons of clean, chlorinated water (see Table 5) to the well to force the solution out into the formation. One-half teaspoon of calcium hypochlorite or one-half cup of laundry bleach in 5 gallons of water is enough for this purpose.

4. Wash the exterior surface of the pump cylinder and drop pipe as they are lowered into the well.

5. After the pump has been set in position, operate the pump until a distinct odor of chlorine can be detected in the water discharged.

6. Allow the chlorine solution to remain in the well for at least 4 hours – preferably overnight.

7. After disinfection, pump the well until the odor of chlorine can no longer be noticed in the water discharged.

In the case of deep wells having a high water level, it may be necessary to resort to special methods of introducing the disinfecting agent into the well so as to insure proper diffusion of chlorine throughout the well. The following method is suggested.

Place the granulated calcium hypochlorite in a short section of pipe capped at both ends. A number of small holes should be drilled through each cap or into the sides of the pipe. One of the caps should be fitted with an eye to facilitate attachment of a suitable cable. The disinfecting agent is distributed when the pipe section is lowered or raised throughout the depth of the water.

Flowing Artesian Wells

The water from flowing artesian wells is generally free from contamination as soon as the well is completed or after it has been allowed to flow a short time. It is therefore not generally necessary to disinfect flowing wells. If, however, analyses show persistent contamination, the well should be thoroughly disinfected as follows.

Use a device such as the pipe described in the preceding section or any other appropriate device by means of which a surplus supply of disinfectant can be placed at or near the bottom of the well. The cable supporting the device can be passed through a stuffing box at the top of the well. After the disinfectant has been placed at or near the bottom of the well, throttle down the flow sufficiently to obtain an adequate concentration. When water showing an adequate disinfectant concentration appears at the surface, close the valve completely and keep it closed for at least 24 hours.

Bacteriological Tests Following Disinfection

If the bacteriological examination of water samples collected after disinfection indicates that the water is not safe for use, disinfection should be repeated until tests show that water samples from that portion of the system being disinfected are satisfactory. Samples collected immediately after disinfection may not be representative of the water served normally. Hence, if bacteriological samples are collected immediately after disinfection, it is necessary that the sampling be repeated several days later to check on the delivered water under normal conditions of operation and use. The water from the system should not be used for domestic and culinary purposes until the reports of the tests indicate that the water is safe for such uses. If after repeated disinfection the water is unsatisfactory, treatment of the supply is needed to provide water which always meets the Public Health Service Drinking Water Standards. Under these conditions, the supply should not be used for drinking and culinary purposes until adequate treatment has been provided.

Abandonment of Wells

Unsealed, abandoned wells constitute a potential hazard to the public health and welfare of the surrounding area. The sealing of an abandoned well presents certain problems, the solution of which involves consideration of the construction of the well and the geological and hydrological conditions of the area. In the proper sealing of a well, the main factors to be considered are elimination of any physical hazard, the prevention of any possible contamination of the ground water, the conservation and maintenance of the yield and hydrostatic pressure of the aquifer, and the prevention of any possible contact between desirable and undesirable waters.

The basic concept behind the proper sealing of any abandoned well is that of restoration, as far as feasible, of the controlling geological conditions that existed before the well was drilled or constructed. If this restoration can be properly accomplished, an abandoned well will not create a physical or health hazard.

When a well is to be permanently abandoned, the lower portion of it is best protected when filled with concrete, cement grout, neat cement, or clays with sealing properties similar to those of cement. When dug or bored wells are filled, as much of the lining should be removed as possible so that surface water will not reach the water-bearing strata through a porous lining or one containing cracks or fissures. When any question arises, follow the regulations and recommendations of the State or local health department.

Abandoned wells should *never* be used for the disposal of sewage or other wastes.

Reconstruction of Existing Dug Wells

Existing wells used for domestic water supplies subject to contamination should be reconstructed so as to insure safe water. When reconstruction is not practicable, the water supply should be treated or a new well constructed.

Dug wells with stone or brick casings can often be rebuilt by enclosing existing casings with concrete or by the use of a buried concrete slab.

Care must be exercised on entering wells because until properly ventilated they may contain dangerous gases or lack oxygen.

Hydrogen sulfide gas is found in certain ground waters and, being heavier than air, tends to accumulate in excavations. It is explosive, and nearly as poisonous as cyanide. Also, a person's sense of smell tires quickly in its presence, so that he is unable to sense danger. Concentrations may become dangerous without further warning

Methane gas is also found in some ground water or in underground formations. It is the product of the decomposition of organic matter. It is not toxic, but is highly explosive

Gasoline or carbide lanterns or candles may *not* be reliable indicators of safe atmospheres within a well, as many of these devices will continue to burn at oxygen levels well below those safe for humans. Also, any open flame carries the additional risk of an explosion from accumulated combustible gases.

The "flame safety lamp" used by miners, construction companies, and utilities service departments is a much better device for determining safe atmospheres. It is readily obtainable from mine safety equipment suppliers It should be lowered on a rope to the bottom of the well to test the atmosphere. Even after the well has passed this test, the first person to enter the well should carry a safety rope tied around his waist, with two persons standing by, above ground, to rescue him at the first sign of dizziness or other distress.

Improvements should be planned so that the reconstructed well will conform as nearly as possible with the principles set forth in this manual. If there is any doubt as to what should be done, advice should be obtained from the State or local health department

Special Considerations in Constructing Artesian Wells

In order that the water may be conserved and the productivity of an artesian well improved, it is essential that the casing be sealed into the confining stratum. Otherwise, a loss of water may occur by leakage into lower pressure permeable strata at higher elevations A flowing artesian well should be designed so that the movement of water from the aquifer can be controlled Water can be conserved if such a well is equipped with a valve or shutoff device. When the recharge area and aquifer are large and the number of wells which penetrate the aquifer are small, the flowing artesian well produces a fairly steady flow of water throughout the year.

DEVELOPMENT OF SPRINGS

There are two general requirements necessary in the development of a spring used as a source of domestic water: (1) selection of a spring with adequate capacity to provide the required quantity or quality of water for its intended use throughout the year, (2) protection of the sanitary quality of the spring. The measures taken to develop a spring must be tailored to its geological conditions and sources.

The features of a spring encasement are the following: (1) an open-bottom, watertight basin intercepting the source which extends to bedrock or a system of collection pipes and a storage tank, (2) a cover that prevents the entrance of surface drainage or debris into the storage tank, (3) provision for the cleanout and emptying of the tank contents, (4) provision for overflow, and (5) a connection to the distribution system or auxiliary supply. (See fig. 10.)

A tank is usually constructed in place with reinforced concrete of such dimensions as to enclose or intercept as much of the spring as possible When a spring is located on a hillside, the downhill wall and sides are extended to bedrock or to a depth that will insure maintenance of an adequate water level in the tank. Supplementary cutoff walls of concrete or impermeable clay extending laterally from the tank may be used to assist in controlling the water table in the locality of the tank. The lower portion of the uphill wall of the tank can be constructed of stone, brick, or other material so placed that water may move freely into the tank from the formation. Backfill of graded gravel and sand will aid in restricting movement of fine material from the formation toward the tank.

The tank cover should be cast in place to insure a good fit. Forms should be designed to allow for shrinkage of concrete and expansion of form lumber. The cover should extend down over the top edge of the tank at least 2 inches The tank cover should be heavy enough so that it cannot be dislodged by children and should be equipped for locking.

A drain pipe with an exterior valve should be placed close to the wall of the tank near the bottom The pipe should extend horizontally so as to clear the normal ground level at the point of discharge by at least 6 inches. The discharge end of the pipe should be screened to prevent the entrance of rodents and insects.

The overflow is usually placed slightly below the maximum water-level elevation and screened A drain apron of rock should be

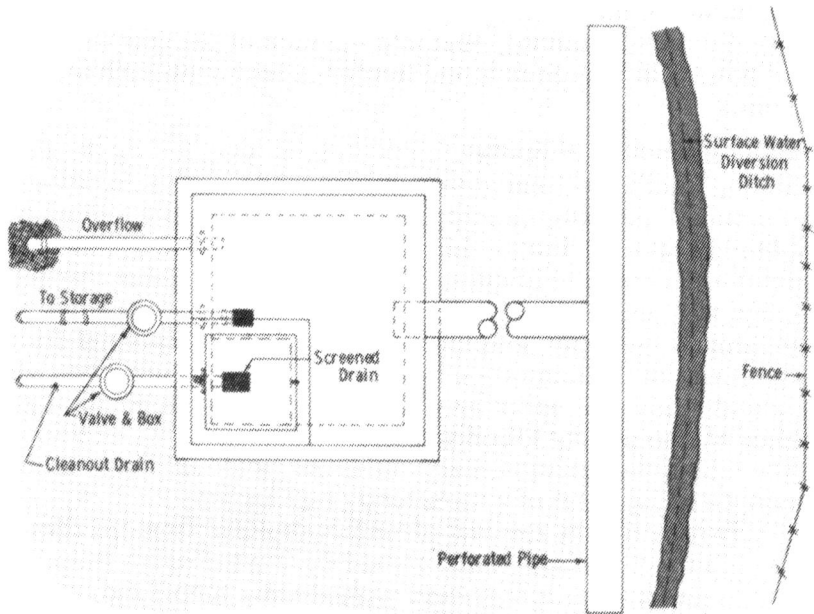

PLAN

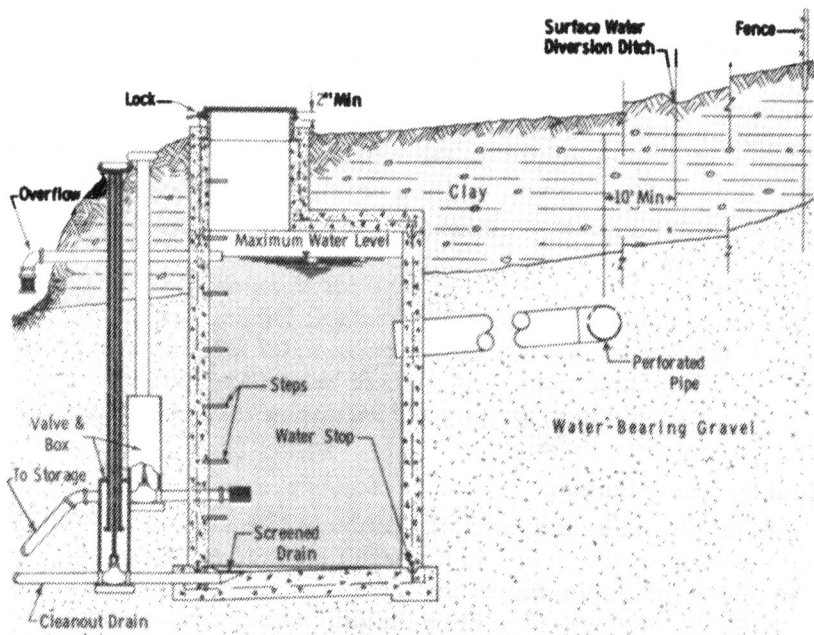

ELEVATION

FIGURE 10. Spring protection.

provided to prevent soil erosion at the point of overflow discharge.

The supply outlet from the developed spring should be located about 6 inches above the drain outlet and properly screened. Care should be taken in casting pipes into the walls of the tank to insure good bond with the concrete and freedom from honeycomb around the pipes.

Sanitary Protection of Springs

Springs usually become contaminated when barnyards, sewers, septic tanks, cesspools, or other sources of pollution are located on higher adjacent land. In limestone formations, however, contaminated material frequently enters the water-bearing channels through sink holes or other large openings and may be carried along with ground water for long distances. Similarly, if material from such sources of contamination finds access to the tubular channels in glacial drift, this water may retain its contamination for long periods of time and for long distances.

The following precautionary measures will help to insure developed spring water of a consistently high quality:

1. Provide for the removal of surface drainage from the site. A surface drainage ditch should be located uphill from the source so as to intercept surface-water runoff and carry it away from the source. Location of the ditch and the points at which the water should be discharged are a matter of judgment. Criteria used should include the topography, the subsurface geology, land ownership, and land use.

2. Construct a fence to prevent entry of livestock. Its location should be guided by the considerations mentioned in item 1. The fence should exclude livestock from the surface-water drainage system at all points uphill from the source.

3. Provide for access to the tank for maintenance, but prevent removal of the cover by a suitable locking device.

4. Monitor the quality of the spring water with periodic checks for contamination. A marked increase in turbidity or flow after a rainstorm is a good indication that surface runoff is reaching the spring.

Disinfection of Springs

Spring encasements should be disinfected by a procedure similar to that used for dug wells. If the water pressure is not sufficient to raise the water to the top of the encasement, it may be possible to shut off the flow and thus keep the disinfectant in the encasement for 24 hours. If the flow cannot be shut off entirely, arrangements should be made to supply disinfectant continuously for as long a period as practicable.

INFILTRATION GALLERIES

Many recreational or other developments located in the mountains have access to water supplies that are located near the headwaters of mountain streams where the watersheds are generally heavily forested and uninhabited by man. Even under these conditions, pathogenic bacteria – in addition to soil bacteria – have been found in the water.

Some of the major problems which are encountered in operating and maintaining these supplies are created by debris and turbidity encountered at the waterworks intake following spring thaws and periods of heavy rainfall. When practical, arrangements should be made to remove this material before it reaches the intake. Experience has demonstrated that this material can be removed successfully, especially when small volumes of water are involved, by installing an infiltration gallery at or near the intake.

Where soil formations adjoining a stream are favorable, the water can be intercepted by an infiltration gallery located a reasonable distance from the high-water level and a safe distance below the ground surface. The gallery should be installed so that it will intercept the flow from the stream after flowing through the intervening soil formations between the stream and infiltration gallery.

A typical installation generally involves the construction of an underdrained, sand-filter trench located parallel to the stream bed and about 10 feet from the high-water mark. The sand filter is usually located in a trench with a minimum width of 30 inches and a depth of about 10 feet. At the bottom of the trench, perforated or open joint tile is laid in a bed of gravel about 12 inches in thickness with about 4 inches of graded gravel located over the top of the tile to support the filtering material. The embedded tile is then covered with clean, coarse sand to a minimum depth of 24 inches, and the remainder of the trench backfilled with fairly impervious material. The collection tile is terminated in a watertight, concrete basin from where it is diverted or pumped to the distribution system following chlorination.

Where soil formations adjoining a stream are unfavorable for the location of an infiltration gallery, the debris and turbidity which are occasionally encountered in a mountain stream can be removed by constructing a modified infiltration gallery-slow sand filter combination in the stream bed. A typical installation involves the construction of a dam across the stream to form a natural pool or the excavation of a pool behind the dam. The filter is installed in the pool behind the dam by laying perforated pipe in a bed of graded gravel which is covered by at least 24 inches of clean, coarse sand. About 24 inches of free board should be allowed between the surface of the sand and the dam spillway. The collection lines usually terminate in a watertight, concrete basin located adjacent to the upstream face of the dam from where the water is diverted to chlorination facilities. Experience with these units indicates that they provide satisfactory service with limited maintenance.

Part III

Surface Water for Rural Use

The selection and use of surface-water sources for individual water supply systems require consideration of additional factors not usually associated with ground water sources. When small streams, open ponds, lakes, or open reservoirs must be used as sources of water supply, the danger of contamination and of the consequent spread of enteric diseases such as typhoid fever and dysentery is increased. As a rule, surface water should be used only when ground water sources are not available or are inadequate. Clear water is not always safe, and the old saying that running water "purifies itself" to drinking water quality within a stated distance is false.

The physical and bacteriological contamination of surface water makes it necessary to regard such sources of supply as unsafe for domestic use unless reliable treatment, including filtration and disinfection, is provided.

The treatment of surface water to insure a constant, safe supply requires diligent attention to operation and maintenance by the owner of the system.

When ground water sources are limited, consideration should be given to their development for domestic purposes only. Surface-water sources can then provide water needed for stock and poultry watering, gardening, firefighting, and similar purposes. Treatment of surface water used for livestock is not generally considered essential. There is, however, a trend to provide stock and poultry drinking water which is free from bacterial contamination and certain chemical elements.

SOURCES OF SURFACE WATER

Principal sources of surface water which may be developed include controlled catchments, ponds or lakes, surface streams, and irrigation canals. Except for irrigation canals, where discharges are dependent on irrigation activity, these sources derive water from direct precipitation over the drainage area.

Because of the complexities of the hydrological, geological, and meteorological factors affecting surface-water sources, it is recommended that in planning the development of natural catchment areas of more than a few acres, engineering advice be obtained.

To estimate the yield of the source, it is necessary for one to consider the following information pertaining to the drainage area.

1. Total annual precipitation.
2. Seasonal distribution of precipitation.
3. Annual or monthly variations of rainfall from normal levels.
4. Annual and monthly evaporation and transpiration rates.
5. Soil moisture requirements and infiltration rates.
6. Runoff gage information.
7. All available local experience records.

Much of the required data, particularly that concerning precipitation, can be obtained from publications of the U.S. Weather Bureau. Essential data such as soil moisture and evapotranspiration requirements may be obtained from local soil conservation and agricultural agencies or from field tests conducted by hydrologists.

Controlled Catchments

In some areas ground water is so inaccessible or so highly mineralized that it is not satisfactory for domestic use. In these cases the use of controlled catchments and cisterns may be necessary. A properly located and constructed controlled catchment and cistern, augmented by a satisfactory filtration unit and adequate disinfection facilities, will provide a safe water.

A controlled catchment is a defined surface area from which rainfall runoff is collected. It may be a roof or a paved ground surface. The collected water is stored in a constructed covered tank called a cistern or reservoir. Ground-surface catchments should be fenced to prevent unauthorized entrance by either man or animals. There should be no possibility of the mixture of undesirable surface drainage and controlled runoff. An intercepting drainage ditch around the upper edge of the area and a raised curb around the surface will prevent the entry of any undesirable surface drainage.

For these controlled catchments, simple guidelines to determine water yield from rainfall totals can be established. When the controlled catchment area has a smooth surface or is paved and the runoff is collected in a cistern, water loss due to evaporation, replacement of soil moisture deficit, and infiltration is small. As a general rule, losses from smooth concrete or asphalt-covered ground catchments average less than 10 percent; for shingled roofs or tar and gravel surfaces losses should not exceed 15 percent, and for sheet metal roofs the loss is negligible.

A conservative design can be based on the assumption that the amount of water that can be recovered for use is three-fourths of the total annual rainfall. (See fig. 11.)

Location A controlled catchment may be suitably located on a hillside near the edge of a natural bench. The catchment area can be placed on a moderate slope above the receiving cistern.

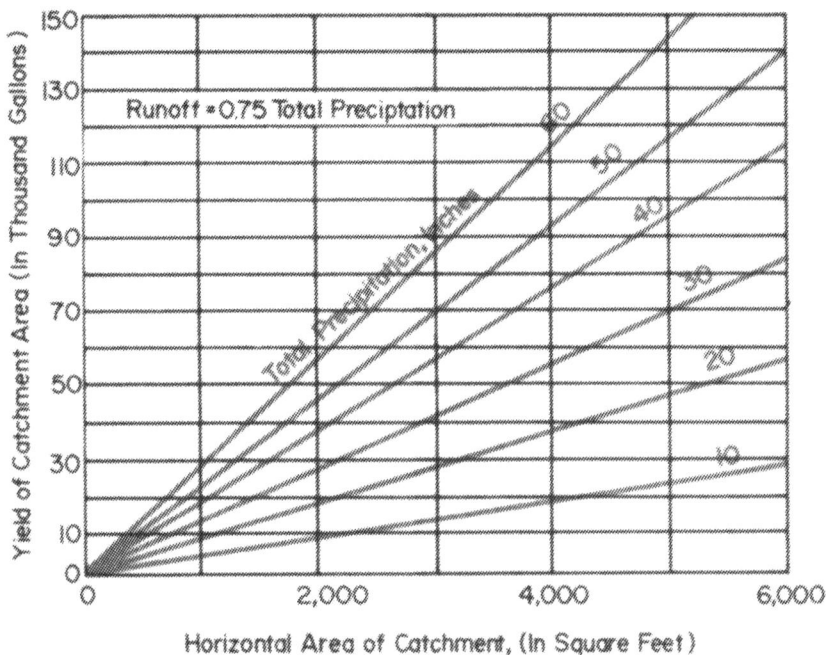

FIGURE 11. Yield of impervious catchment area.

The location of the cistern should be governed by both convenience and quality protection. A cistern should be as close to the point of ultimate use as practical. A cistern should not be placed closer than 50 feet from any part of a sewage-disposal installation, and should be on higher ground.

Cisterns collecting water from roof surfaces should be located adjacent to the building, but not in basements subject to flooding. They may be placed below the surface of the ground for protection against freezing in cold climates and to keep water temperatures low in warm climates but should be situated on the highest ground practicable, with the surrounding area graded to provide good drainage.

Size. The size of a cistern needed will depend on the size of the family and the length of time between periods of heavy rainfall. Daily water requirements can be estimated from Table 1, p. 15. The size of the catchment or roof will depend on the amount of rainfall and the character of the surface. It is desirable to allow a safety factor for lower than normal rainfall levels. Designing for two-thirds of the mean annual rainfall will result usually in a catchment area of adequate capacity.

The following example illustrates the procedure for determining the size of the cistern and required catchment area. Assume that the minimum drinking and culinary requirements of a family of four persons are 100 gallons per day[1] (4 persons × 25 gallons per day = 100 gallons) and that the effective period[2] between rainy periods is 150 days. The minimum volume of the cistern required will be 15,000 gallons (100×150). This volume could be held by a cistern 10 feet deep and 15 feet square. If the mean annual rainfall is 50 inches, then the total design rainfall is 33 inches (50×2/3). In figure 11 the catchment area required to produce 36,500 gallons (365 days × 100 gallons per day) — the total year's requirement — is 2,400 square feet.

Construction. Cisterns should be of watertight construction with smooth interior surfaces. Manhole or other covers should be tight to prevent the entrance of light, dust, surface water, insects, and animals.

Manhole openings should have a watertight curb with edges projecting a minimum of 4 inches above the level of the surrounding surface. The edges of the manhole cover should overlap the curb and project downward a minimum of 2 inches. The covers should be provided with locks to minimize the danger of contamination and accidents.

Provision can be made for diverting initial runoff from paved surfaces or roof tops before the runoff is allowed to enter the cistern. (See fig. 12.)

Inlet, outlet, and waste pipes should be effectively screened. Cistern drains and waste or sewer lines should not be connected.

Underground cisterns can be built of brick or stone, although reinforced concrete is preferable. If used, brick or stone must be low in permeability and laid with full portland cement mortar joints. Brick should be wet before laying. *High-quality workmanship is required, and the use of unskilled labor for laying brick or stone is not advisable.* Two 1/2-inch plaster coats of 1:3 portland cement mortar on the interior surface will aid in providing waterproofing. A hard impervious surface can be made by troweling the final coat before it is fully hardened.

Figure 12 shows a suggested design for a cistern of reinforced concrete. A dense concrete should be used to obtain watertightness and should be vibrated adequately during construction to eliminate honeycomb. All masonry cisterns should be allowed to wet cure properly before being used.

The procedures outlined in part V of this manual should be followed in disinfecting the cistern with chlorine solutions. Initial

---

[1] Twenty-five gallons per person per day, assuming that other uses are supplied by water of poorer quality

[2] Effective period is the number of days between periods of rainfall during which there is negligible precipitation.

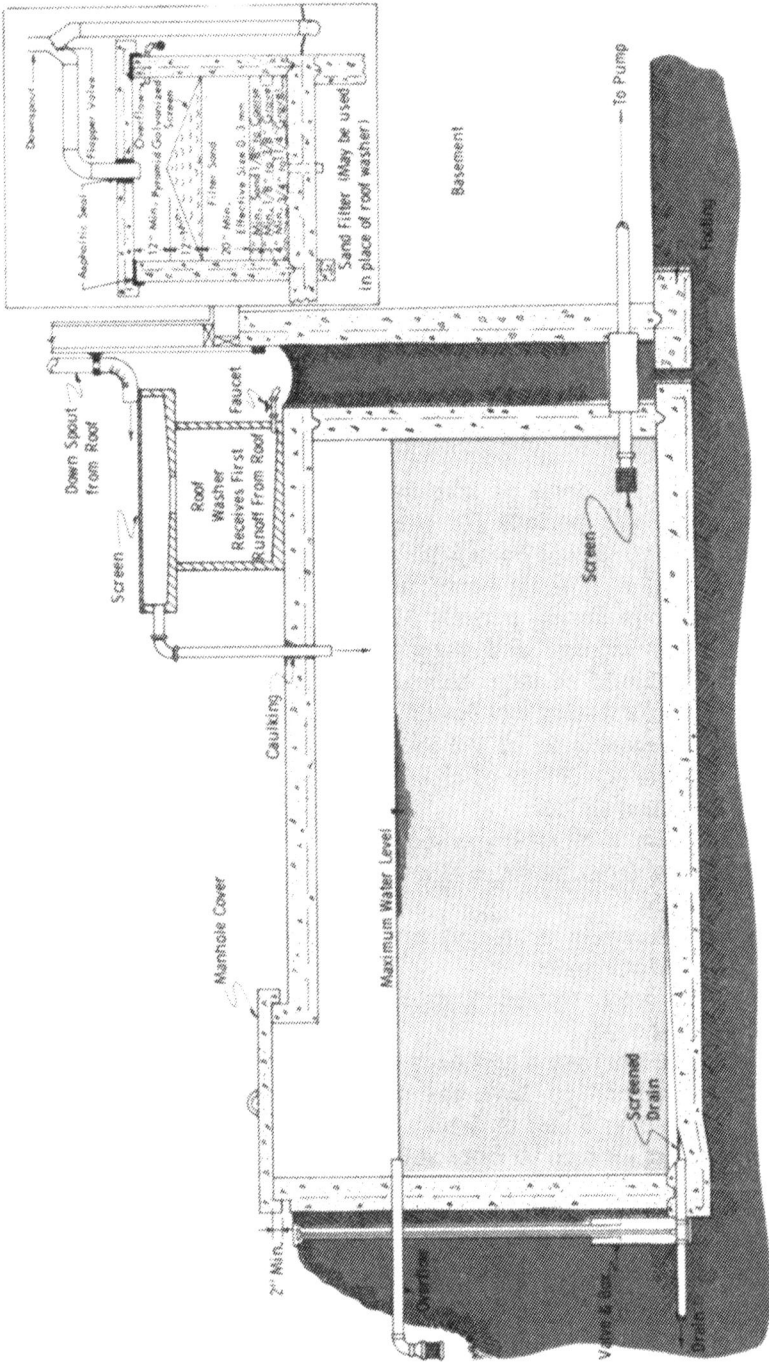

FIGURE 12. Cistern.

and periodic water samples should be taken to determine the bacteriological quality of the water supply. Chlorination may be required on a continuing basis if the bacteriological results indicate that the quality is unsatisfactory.

Ponds or Lakes

A pond or lake should be considered as a source of water supply only after ground water sources and controlled catchment systems are found to be inadequate or unacceptable. The development of a pond as a supply source depends on several factors: (1) the selection of a watershed that permits only water of the highest quality to enter the pond, (2) usage of the best water collected in the pond, (3) filtration of the water to remove turbidity and reduce bacteria, (4) disinfection of filtered water, (5) proper storage of the treated water, and (6) proper maintenance of the entire water system. Local authorities may be able to furnish advice on pond development.

The value of a pond or lake development is its ability to store water during wet periods for use during periods of little or no rainfall. A pond should be capable of storing a minimum of 1 year's supply of water. It must be of sufficient capacity to meet water supply demands during periods of low rainfall with an additional allowance for seepage and evaporation losses. The drainage area (watershed) should be large enough to catch sufficient water to fill the pond or lake during wet seasons of the year.

Careful consideration of the location of the watershed and pond site reduces the possibility of chance contamination.

The watershed should:
1. Be clean, preferably grassed.
2. Be free from barns, septic tanks, privies, and soil-absorption fields.
3. Be effectively protected against erosion and drainage from livestock areas.
4. Be fenced to exclude livestock.

The pond should:
1. Be not less than 8 feet deep at deepest point.
2. Be designed to have the maximum possible water storage area over 3 feet in depth.
3. Be large enough to store at least 1 year's supply.
4. Be fenced to keep out livestock.
5. Be kept free of weeds, algae, and floating debris.

In many instances pond development requires the construction of an embankment with an overflow or spillway. Assistance in designing a storage pond may be available from Federal, State, or local health agencies; the U.S. Soil Conservation Service; and in publications from the State or county agricultural, geological, or soil conservation departments. For specific conditions, engineering or geological advice may be needed.

Intake. A pond intake must be properly located in order that it may draw water of the highest possible quality. When the intake is placed too close to the pond bottom, it may draw turbid water or water containing decayed organic material. When placed too near the pond surface, the intake system may draw floating debris, algae, and aquatic plants. The depth at which it operates best will vary, depending upon the season of the year and the layout of the pond. The most desirable water is usually obtained when the intake is located between 12 and 18 inches below the water surface. An intake located at the deepest point of the pond makes maximum use of stored water.

Pond intakes should be of the type illustrated in figure 13. This is known as a floating intake. The intake consists of a flexible pipe attached to a rigid conduit which passes through the pond embankment.

In accordance with applicable specifications, gate valves should be installed on the main line below the dam and on any branch line to facilitate control of the rate of discharge.

Treatment The pond water-treatment facility consists of four general parts. (See fig. 14.)

1. *Settling Basin.* The first unit is a settling basin. The purpose of the basin is to allow the large particles of turbidity to settle. This may be adequately accomplished in the pond. When this is not completely effective, a properly designed settling basin with provision for coagulation may be needed. The turbid water is mixed with a suitable chemical such as alum. Alum and other chemical aids speed up the settling rate of suspended materials present in the water. This initial process helps to reduce the turbidity of the water to be passed through the filter. Addition of alum will lower the pH, which may have to be readjusted with lime if corrosion of the distribution piping results.

2. *Filtration Unit.* After settling, the water moves to a second compartment where it passes through a filter bed of sand and gravel. The suspended particles which are not removed by settlement or flocculation are now removed.

3. *Clear Water Storage.* After the water leaves the filter, it drains into a clear well, cistern, or storage tank.

4. *Disinfection.* After water has settled and has been filtered it must be disinfected. Proper disinfection is the most important part of pond-water treatment. The continuous operation and high-quality performance of the equipment are very important. The different types of equipment and processes are described in detail in part IV. When the water is chlorinated, livestock unaccustomed to chlorinated water may refuse to drink the water for several days.

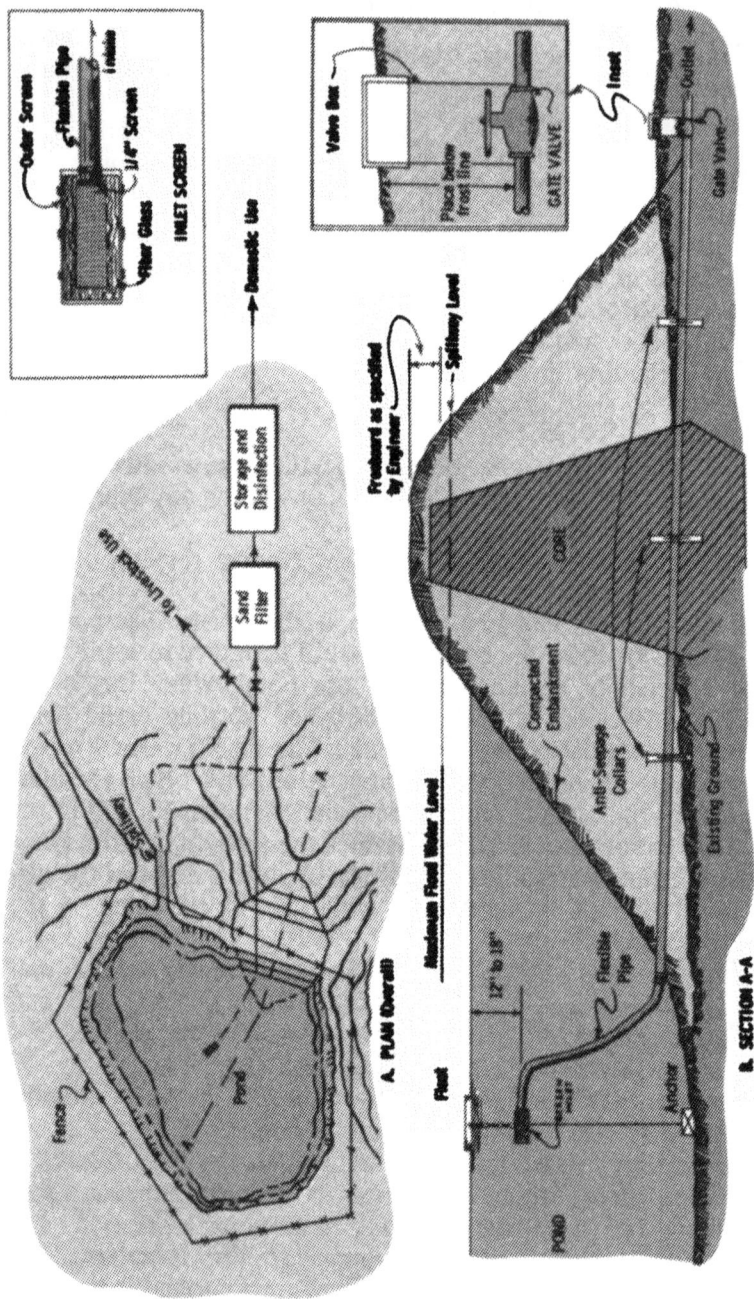

FIGURE 13. Pond.

68

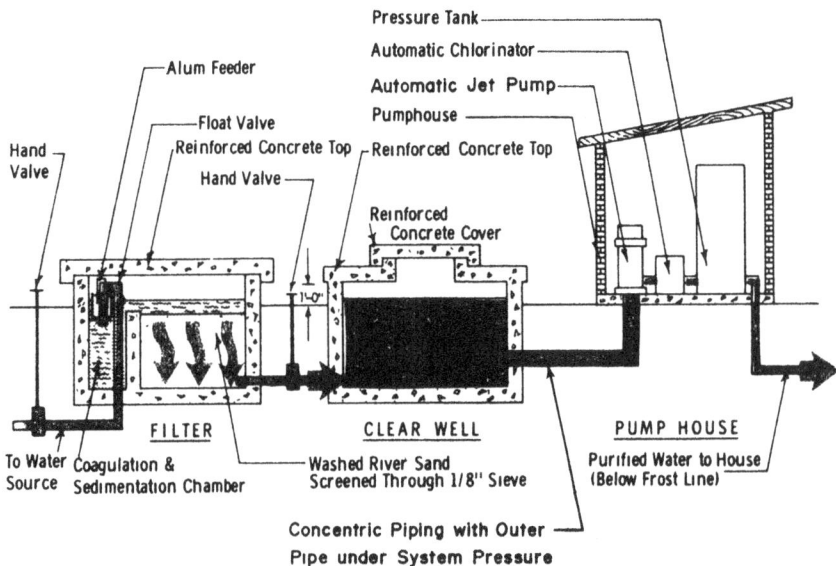

FIGURE 14. Schematic diagram of pond water-treatment system.

They usually become accustomed to it within a short period of time.

Bacteriological Examination. After the treatment and disinfection equipment have been checked and are operating satisfactorily, a bacteriological examination of a water sample should be made. Before a sample is collected, the examining laboratory should be contacted for its recommendations. These recommendations should include the type of container to be used and the method and precautions to take during collection, handling, and mailing. When no other recommendations are available, follow those given in appendix B.

Water should *not* be used for drinking and culinary purposes until the results of the bacteriological examination show the water to be safe.

The frequency of subsequent bacteriological examinations should be based on any breakdown or changes made in the sanitary construction or protective measures associated with the supply. A daily determination and record of the chlorine residual is recommended to insure that proper disinfection is accomplished.

Plant Maintenance. The treatment facility should be inspected daily. The disinfection equipment should be checked to make sure it is operating satisfactorily. When chlorine disinfection is practiced, the chlorinator and the supply of chlorine solution should be checked. The water supply should be checked daily for its chlorine residual. The water may become turbid after heavy rains and the quality may change. Increases in the amount of chlorine and coagulants used will then be required. The performance of the filter should be watched closely. When the water becomes turbid or the available quantity of water decreases, the filter should be cleaned or backwashed.

Protection From Freezing Protection against freezing must be provided unless the plant is not operated and is drained during freezing weather. In general, the filter and pumproom should be located in a building that can be heated in winter. With suitable topography the need for heat can be eliminated by placement of the pumproom and filter underground on a hillside. Gravity drainage from the pumproom must be possible to prevent flooding. No matter what the arrangement, the filter and pumproom must be easily accessible for maintenance and operation.

Tastes and Odors. Surface water frequently develops musty or undesirable tastes and odors. These are generally caused by the presence of microscopic plants called algae. There are many kinds of algae. Some occur in long threadlike filaments that are visible as large green masses of scum; others may be separately free floating and entirely invisible to the unaided eye. Some varieties may grow in great quantities in the early spring, others in summer, and still others in the fall. Tastes and odors generally result from the decay of dead algae. This decay occurs naturally as plants pass through their life cycle. For additional discussion, see "Control of Algae" in part IV.

Tastes and odors in water can usually be satisfactorily removed by passing the previously filtered and chlorinated surface water through an activated carbon filter. These filters may be helpful in improving the taste of small quantities of previously treated water used for drinking or culinary purposes. They also absorb excess chlorine. Carbon filters are commercially available, and require periodic servicing.

Carbon filters should not be expected to be a substitute for sand filtration and disinfection. They have insufficient area to handle raw surface water and will clog very rapidly when filtering turbid water.

Weed Control. The growth of weeds around a pond should be controlled by cutting or pulling. Before weedkillers are used, the local health department should be contacted for advice since herbicides often contain compounds that are highly toxic to humans and animals. Algae in the pond should be controlled, particularly the blue-green types that produce scum and objectionable odors and that, in unusual instances, may harm livestock. (See pt. IV.)

Streams

Streams receiving runoff from large uncontrolled watersheds may be the only source of water supply. The physical and bacteriological quality of surface water varies and may impose unusually or abnormally high loads on the treatment facilities.

Stream intakes should be located upstream from sewer outlets or other sources of contamination. The water should be pumped when the silt load is low. A low-water stage usually means that the temperature of the water is higher than normal and the water is of poor chemical quality. Maximum silt loads, however, occur during maximum runoff. High-water stages shortly after storms are usually the most favorable for diverting or pumping water to storage. These conditions vary and should be determined for the particular stream.

Irrigation Canals

If properly treated, irrigation water may be used as a source of domestic water supply. Water obtained from irrigation canals should be treated the same as water from any other surface-water source. For additional information, see part IV.

When return irrigation (tail water) is practiced, the water may contain large concentrations of undesirable chemicals, including pesticides, herbicides, and fertilizer. Whenever water from return irrigation is used for domestic purposes, a periodic chemical analysis should be made. Because of the poor quality of this water, it should only be used if no other water source is available.

Part IV

Water Treatment

NEED AND PURPOSE

Raw waters obtained from natural sources may not be completely satisfactory for domestic use. Surface waters may contain pathogenic (disease-producing) organisms, suspended matter, or organic substances. Except in limestone areas, ground water is less likely to have pathogenic organisms than surface water, but may contain undesirable tastes and odors or mineral impurities limiting its use or acceptability. Some of these objectionable characteristics may be tolerated temporarily, but it is desirable to raise the quality of the water to the highest possible level by suitable treatment. In those instances where the nearly ideal water can be developed from a source, it is still advisable to provide the necessary equipment for treatment to insure safe water at all times.

The quality of surface water constantly changes. Natural processes which affect water quality are the dissolving of minerals, sedimentation, filtration, aeration, sunlight, and biochemical decomposition. Natural processes may tend to pollute and contaminate or to purify the water; however, the natural processes of purification are not consistent or reliable.

Bacteria which are numerous in waters at or near the earth's surface may be reduced by soil filtration, depletion of available oxygen, or underground detention for long periods under conditions unfavorable for bacterial growth or survival. When water flows through underground fissures or channels, however, it may retain contamination over long distances and for extended periods of time.

The false belief that flowing water purifies itself after traveling various distances has led to unjustified feelings of security about its safety. Under certain conditions the number of micro-organisms in flowing surface water may increase instead of decrease.

Water treatment incorporates, modifies, or supplements certain natural processes. This provides adequate assurance that the water is free from pathogenic organisms or other undesirable materials or chemicals. Water treatment may condition or reduce to acceptable levels any chemicals or esthetically objectionable impurities which may be present in the water.

Some of the natural treatment processes and manmade

adaptations to improve and condition water are discussed in the following sections.

SEDIMENTATION

Sedimentation is a process of gravity settling and deposition of comparatively heavy suspended material in water.

This settling action can be accomplished in a quiescent pond or properly constructed tank or basin. At least 24 hours' detention time must be provided if a significant reduction in suspended matter is to be accomplished. The inlet of the tank should be arranged so that the incoming water containing suspended matter is distributed uniformly across the entire width as the water flows to the outlet located at the opposite end. Baffles are usually constructed to reduce high local velocities and short circuiting of the water. The cleaning and repairing of an installation can be facilitated by the use of a tank designed with two separated sections, each of which may be used independently.

COAGULATION-FLOCCULATION

Coagulation is the process of forming flocculent particles in a liquid by the addition of a chemical. Coagulation is achieved by adding to the water a chemical such as alum (hydrated aluminum sulfate). The chemical is mixed with the turbid water and then allowed to remain quiet. The suspended particles will combine physically and form a floc. The floc or larger particles will settle to the bottom of the basin. This may be done in a separate tank or in the same tank after the mixing has been stopped. Adjustment of pH may be required after sedimentation. Some colors can be removed from water by using proper coagulation techniques. Competent engineering advice, however, should be obtained on specific coagulation problems.

FILTRATION

Filtration is the process of removing suspended matter from water as it passes through beds of porous material. The degree of removal depends on the character and size of the filter media, the thickness of the porous media, and the size and quantity of the suspended solids. Since bacteria can travel long distances through granular materials, filters should not be relied upon to produce bacteriologically safe water, even though they may greatly improve the quality. When a water source contains a large amount of turbidity, a large portion of it can first be removed by sedimentation. A protected pond with gentle grassy slopes is often helpful in producing a reasonably clear raw water. This action will reduce the load placed on the filters.

Types of filters that may be used include:

Slow Sand Filters. Water passes slowly through beds of fine sand at rates averaging 0.05 gallon per minute per square foot of filter area.

Pressure Sand Filters. Water is applied at a rate at or above 2 gallons per minute per square foot of filter area with provisions made for frequent backwashing of the filters.

Diatomaceous Earth Filters. Suspended solids are removed by passing the water through a layer of diatomaceous filter media supported by a rigid base septum at rates approximately that of pressure sand filters.

Porous Stone, Ceramic, or Unglazed Porcelain Filters (Pasteur Filters) These are small household filters that are attached to faucets.

Properly constructed slow sand filters require a minimum of maintenance and can be easily adapted to individual water systems. The length of time between cleaning will vary from a day to a week or month; the length of the interval depends upon the turbidity of the water. After an interval it is necessary to clean the filter by removing approximately 1 inch of sand from the surface of the filter and either discarding it or stockpiling it for subsequent washing and reuse. This removal will necessitate the periodic addition of new or washed sand.

Sand for slow sand filters should consist of hard, durable grains free from clay, loam, dirt, and organic matter. It should have a sieve analysis which falls within the range of values shown in table 6.

TABLE 6. — *Recommended mechanical analysis of slow sand filter media*

Material passing sieve (percent)	U.S. sieve no.	Material passing sieve (percent)	U.S. sieve no
99	4	33-55	30
90-97	12	17-35	40
75-90	16	4-10	60
60-80	20	1	100

Sands with an effective size of 0.20 to 0.40 millimeter are satisfactory. The effective size is the size of the grain in millimeters, such that 10 percent of the material, by weight, is of a smaller size. The uniformity coefficient should be between 2.0 and 3.0. The uniformity coefficient is taken as the ratio of the grain size that has 60 percent finer than itself to the size that has 10 percent finer than itself.

For best results the rate of filtration for a slow sand filter should be 60 to 180 gallons per day per square foot of filter bed surface. The amount of water that flows through the filter bed can be adjusted by a valve placed on the effluent line. Between 27 and 36 inches of sand, with an additional 6 to 12 inches that can be removed during cleaning, is usually sufficient. Six to 8 inches of gravel will support the sand and keep it out of the underdrain system. A 1¼-inch plastic pipe drilled with 3/4-inch holes facing down makes a convenient underdrain system. One to 2 feet of freeboard on the top of the filter is usually sufficient.

Rapid sand filtration is not usually desirable for small individual water supplies because of the necessary controls and additional attention required to obtain satisfactory results. When adequate operation and supervision are provided, pressure sand filtration can be used successfully.

Diatomaceous earth filters, which require periodic attention, are of two types — vacuum or pressure. These filters are effective when properly operated and maintained.

The effectiveness of filtration is monitored by measurement of turbidity, a light-scattering property of particles suspended in water. Filtered water must contain low turbidity if adequate disinfection is to be accomplished.

The possibility must be considered that dirty stone or ceramic faucet filters may attract bacteria and provide a place for their multiplication or that these filters may develop hairline cracks. For these reasons, small household faucet filters cannot be depended upon to remove pathogenic bacteria, and their use is not recommended for producing bacteriologically safe water.

Small pad, spool, or wad coarse filters may be useful for low-capacity, coarse filtration for removal of large suspended particles only. Proper disinfection of water before consumption is necessary to assure its safety.

DISINFECTION

The most important water treatment process is disinfection. Disinfection is necessary to destroy all pathogenic bacteria and other harmful organisms that may be present in water. If complete destruction of these organisms is to be accomplished, the water to which the disinfectant is added must be low in turbidity. After disinfection, water must be kept in suitable tanks or other storage facilities to prevent recontamination.

Chemical Disinfection

The desirable properties for a chemical disinfectant are high germicidal power, stability, solubility, nontoxicity to man or animals, economy, dependability, residual effect, ease of use and measurement, and availability.

Compounds of chlorine most satisfactorily comply with the desirable properties of a chemical disinfectant; and as a result, chlorine is the most commonly used water disinfectant.

Disinfectant Terminology

1. *Chlorine concentration.* This is expressed in milligrams per liter (mg/ℓ). One mg/ℓ is equivalent to 1 milligram of

chlorine in 1 liter of water. For water, the terms parts per million (ppm) and mg/ℓ are essentially equal.

2. *Chlorine feed or dosage.* The actual amount in mg/ℓ fed into the water system by feeder or automatic dosing apparatus is the chlorine feed or dosage.

3. *Chlorine demand.* The chlorine fed into the water that combines with the impurities, and, therefore, may not be available for disinfection action, is commonly called the chlorine demand of the water. Examples of impurities causing chlorine demand are organic materials and certain "reducing" materials such as hydrogen sulfide, ferrous iron, nitrites, etc.

4. *Free and combined chlorine.* In addition to organic materials that exert a chlorine demand, chlorine can combine with ammonia nitrogen, if any is present in the water, to form chlorine compounds that have some biocidal properties. These chlorine compounds are called combined chlorine residual. If no ammonia is present in the water, however, the chlorine that remains in the water once the chlorine demand has been satisfied is called free chlorine residual.

5. *Chlorine contact time.* The chlorine contact time is the period of time that elapses between the time when the chlorine is added to the water and the time when that particular water is used. Contact time is required for chlorine to act as a disinfectant.

Chlorine Disinfection

In general, the primary factors that determine the biocidal efficiency of chlorine are as follows:

1. *Chlorine concentration.* The higher the concentration, the more effective the disinfection and the faster the disinfection rate.

2. *Type of chlorine residual.* Free chlorine is a much more effective disinfectant than combined chlorine.

3. *Contact time between the organism and chlorine.* The longer the time, the more effective the disinfection.

4. *Temperature of the water in which contact is made.* The higher the temperature, the more effective the disinfection.

5. *The pH of the water in which contact is made.* The lower the pH, the more effective the disinfection.

Chlorine dosage should be great enough to satisfy the chlorine demand and provide a residual of 0.4 mg/ℓ after a chlorine contact time of 30 minutes or a combined residual of 2.0 mg/ℓ with a 2-hour contact time. Hypochlorinators pump or inject a chlorine solution into the water, and, when they are properly maintained, provide a reliable method for applying chlorine. Hypochlorinators and chlorine residual test equipment are available from several manufacturers through local dealers.

Chlorine Compounds and Solutions

Compounds of chlorine such as sodium or calcium hypochlorite have excellent disinfecting properties. In small water systems these chlorine compounds are usually added to the water in a solution form.

One of the commonly used forms of chlorine is calcium hypochlorite. It is commercially available in the form of soluble powder or tablets. These compounds are classed as high-test hypochlorites and contain 65 to 75 percent available chlorine by weight. Packed in cans or drums, these compounds are stable and will not deteriorate if properly stored and handled.

Prepared sodium hypochlorite solution is available locally through chemical or swimming pool equipment suppliers. The most common type is household chlorine bleach which has a strength of approximately 5 percent available chlorine by weight. Other sodium hypochlorite solutions vary in strength from 3 to 15 percent available chlorine by weight, and are reasonably stable when stored in a cool, dark place. These solutions are diluted with potable water to obtain the desired solution strength to be fed into the system.

When hypochlorite powders are used, fresh chlorine solutions should be prepared at frequent intervals because the strength of chlorine solutions deteriorates gradually after preparation. The container or vessel used for preparation, storage, or distribution of the chlorine solution should be resistant to corrosion and light. (Light produces a photochemical reaction that reduces the strength of chlorine solutions.) Suitable materials include glass, plastic, crockery, or rubber-lined metal containers.

Hypochlorite solutions are used either full strength as prepared or are diluted to solution strength suited to the feeding equipment and the rate of water flow. In preparing these solutions, one must take into account the chlorine content of the concentrated solution. For example, if 5 gallons of 2 percent solution are to be prepared with a high-test calcium hypochlorite powder or tablet containing 70 percent available chlorine, the high-test hypochlorite would weigh 1.2 pounds.

Pounds of compound required

$$= \frac{\dfrac{\% \text{ strength}}{\text{of solution}} \times \dfrac{\text{gallons solution}}{\text{required}} \times 8.3}{\% \text{ available chlorine in compound}}$$

$$= \frac{2 \times 5 \times 8.3}{70}$$

$$= 1.2 \text{ pounds}$$

Expressed in another way, 1.2 pounds of high-test hypochlorite with 70 percent available chlorine would be added to 5 gallons of water to produce a 2-percent chlorine solution.

Determination of Chlorine Residual

Residual chlorine can exist in water as a chlorine compound of organic matter and ammonia or as both combined and free available chlorine residual. When present as a chlorine compound, it is called combined available chlorine residual, as free chlorine it is known as free available chlorine residual, and as both combined and free available chlorine it is called total available chlorine residual. Thus, "sufficient chlorine" is that amount required to produce a desired residual after a definite contact period, whether combined, free, or total.

The amount of chlorine remaining (chlorine residual) in the water is determined by a relatively simple test commonly called the DPD colorimetric test, short for the chemical name N,N-diethyl-p-phenylene-diamine. The test may be done under "field" conditions, using pill reagents that are placed in a special test tube. The presence of free chlorine residual produces a violet color that can be compared with color standards to determine the quantity present. The kits, complete with all necessary tubes, chemicals, color standards and instructions, can be obtained from firms that specialize in the manufacture of water testing equipment and materials. A combination DPD and pH kit is available; its modest price makes it a good investment. State and county water supply agencies can provide the names of kits they consider acceptable.

Wherever chlorination is required for disinfection, testing for chlorine residual should be done at least daily.

For those desiring more information on the DPD test, a description is included in Standard Methods for the Examination of Water and Wastewater.[1]

---

[1]Obtainable from the American Public Health Association, 1015 Fifteenth Street, NW., Washington, D.C. 20036

Chlorination Equipment

Hypochlorinators

Hypochlorinators pump or inject a chlorine solution into the water. When they are properly maintained, hypochlorinators provide a reliable method for applying chlorine to disinfect water.

Types of hypochlorinators include positive displacement feeders, aspirator feeders, suction feeders, and tablet hypochlorinators.

Positive Displacement Feeders. A common type of positive displacement hypochlorinator is one that uses a piston or diaphragm pump to inject the solution. This type of equipment, which is adjustable during operation, can be designed to give reliable and accurate feed rates. When electricity is available, the stopping and starting of the hypochlorinator can be synchronized with the pumping unit. A hypochlorinator of this kind can be used with any water system; however, it is especially desirable in systems where water pressure is low and fluctuating.

Aspirator Feeders. The aspirator feeder operates on a simple hydraulic principle that employs the use of the vacuum created when water flows either through a venturi tube or perpendicular to a nozzle. The vacuum created draws the chlorine solution from a container into the chlorinator unit where it is mixed with water passing through the unit, and the solution is then injected into the water system. In most cases, the water inlet line to the chlorinator is connected to receive water from the discharge side of the water pump, with the chlorine solution being injected back into the suction side of the same pump. The chlorinator operates only when the pump is operating. Solution flow rate is regulated by means of a control valve, though pressure variations may cause changes in the feed rate.

Suction Feeders. One type of suction feeder consists of a single line that runs from the chlorine solution container through the chlorinator unit and connects to the suction side of the pump. The chlorine solution is pulled from the container by suction created by the operating water pump.

Another type of suction feeder operates on the siphon principle, with the chlorine solution being introduced directly into the well. This type also consists of a single line, but the line terminates in the well below the water surface instead of the influent side of the water pump. When the pump is operating, the chlorinator is activated so that a valve is opened and the chlorine solution is passed into the well.

In each of these units, the solution flow rate is regulated by means of a control valve and the chlorinators operate only when the pump is operating. The pump circuit should be connected to a liquid level control so that the water supply pump operation is interrupted when the chlorine solution is exhausted.

Tablet Hypochlorinators. The tablet hypochlorinating unit consists of a special pot feeder containing calcium hypochlorite tablets. Accurately controlled by means of a flowmeter, small jets of feed water are injected into the lower portion of the tablet bed. The slow dissolution of the tablets provides a continuous source of fresh hypochlorite solution. This unit controls the chlorine solution. This type of chlorinator is used when electricity is not available, but requires adequate maintenance for efficient operation. It can operate where the water pressure is low.

Gaseous Feed Chlorinators

In installations where large quantities of water are treated, chlorine gas in pressure cylinders may be used as the disinfectant. The high cost of this type of chlorination equipment and the safety precautions necessary to guard against accidents do not usually justify its use in individual water supply systems.

Solution Supply Monitor

Sensing units which can be placed in solution containers to sound a warning alarm when the solution goes below a predetermined level are commercially available. This equipment can also be connected to the pump, which will automatically shut off the pump and activate a warning bell. On such a signal the operator will be required to refill the solution container and take necessary steps to insure proper disinfection.

Chlorination Control

As indicated previously, several factors pertaining to a water supply system have a direct bearing on the effectiveness of chlorine. Because of these variable factors, it is not possible to suggest rigid standards of chlorine disinfection applicable to all water supply systems. It is considered desirable, however, to suggest the following practice in this regard for the guidance of persons responsible for water supply operation and maintenance.

Simple Chlorination

Unless bacteriological or other tests indicate the need for maintaining higher minimum concentrations of free residual chlorine, at least 0.4 mg/ℓ of free residual chlorine (see p. 77) should be in contact with the treated water for not less than 30 minutes before the water reaches the first user beyond the point of chlorine application. It is considered desirable to maintain a detectable free chlorine residual at distant points in the distribution system when using simple chlorination; however, the water can be properly disinfected if a minimum contact time of 30 minutes is assured.

A method known as superchlorination-dechlorination is suggested for use in overcoming and simplifying the problem of insufficient contact time in such water systems. By this method chlorine is added to the water in increased amounts (superchlorination) to provide a minimum chlorine residual of 3.0 mg/ℓ for a minimum contact period of 5 minutes. Removal of the excess chlorine (dechlorination) follows to eliminate objectionable chlorine tastes. Dechlorination equipment is commercially available.

Records

Adequate control is also dependent on the maintenance of accurate operating records of the chlorination process. The record should serve as an indicator that proper chlorination is being accomplished and as a guide in improving operations. The record should show the amount of water treated, amount of chlorine used, setting of the chlorinator, time and location of tests, and results of chlorine residual determinations. This information should be kept current and posted near the chlorinator.

Disinfection With Ultraviolet Light

Ultraviolet (UV) light produced from UV lamps has been shown to be an effective bactericide for both air and water. In disinfecting water, the quantity of radiation required is dependent on such factors as turbidity, color, and dissolved iron salts, which adversely affect the penetration of ultraviolet energy through the water. UV light would not be satisfactory for disinfecting water with high turbidity.

Cylindrical units with standard plumbing fittings have been designed for use in waterlines. They should be checked frequently for light intensity and cleaned of any material that would block radiation from reaching the water. A disadvantage of UV light is that it does not provide a residual in the water as does chlorine. Thus, there is no barrier against recontamination in UV-disinfected water. Also, an uninterrupted source of electric power is needed for UV units. The counsel of the State health authority should be obtained before selecting a particular unit for installation.

Other Methods and Materials for Water Disinfection

A number of other materials and methods are used for disinfecting water. Some of these are as follows:

1. Organic chlorine-yielding compounds
2. Bromine
3. Iodine and iodine-yielding organics
4. Ozone
5. Hydrogen peroxide and peroxide-generating compounds
6. Silver
7. Nontoxic organic acids
8. Lime and mild alkaline agents
9. Ultrasonic cavitation
10. Heat treatment

Some of these are old processes on which detailed studies have been made; others are relatively new.

When a question of specific application arises, the recommendations of the State or local health department should be followed.

CONDITIONING

Iron and/or Manganese

The presence of iron and/or manganese in water creates a problem common to many individual water supply systems. When both are present beyond applicable drinking water standards, special attention should be given. Their removal or elimination depends somewhat on type and quantity, and this influences the procedure and possibly the equipment to be used.

Well water is usually clear and colorless when drawn from the faucet or tap. When water containing colorless, dissolved iron is allowed to stand in a cooking container or comes in contact with a sink or bathtub, the iron combines with oxygen from the air to form a reddish-brown precipitate commonly called rust. Manganese acts in a similar manner, but forms a brownish-black precipitate.

These impurities can impart a metallic taste to the water or to any food in whose preparation such a supply is used. Deposits of iron and manganese produce rusty or brown stains on plumbing fixtures, fabrics, dishes, or utensils. The use of soaps or detergents will not remove these stains, and bleaches and alkaline builders (often sodium phosphate) can intensify the staining. After a prolonged period, iron deposits can build up in pressure tanks, water heaters, and pipelines. This buildup reduces the available quantity and pressure of the water supply.

Iron and manganese can be removed by a combination of automatic chlorination and fine filtration. The chlorine chemically oxidizes the iron or manganese (forming a precipitate), kills iron bacteria, and eliminates any disease bacteria which may be present. The fine filter then removes the iron or manganese precipitates. Other techniques, such as aeration followed by filtration, ion exchange with greensand, or treatment with potassium permanganate followed by filtration, will also remove these materials.

Some filters may dechlorinate also. This chlorination-filtration method provides complete correction of such problems and assures disinfection as well.

Insoluble iron or manganese and iron bacteria will intensely "foul" the mineral bed and the valving of a water softener. It is best, therefore, to remove iron and manganese before the water reaches the softener.

When a backwash filter medium is used it is essential that an adequate quantity of water at sufficient pressure be provided for removing the iron precipitate.

Iron Bacteria

Under certain conditions the removal of iron compounds from a water supply may be complicated by the presence of iron bacteria. When dissolved iron and oxygen are present in the water, these bacteria derive the energy they need for their life processes from the oxidation of the iron to its insoluble form. These products accumulate within a gelatinous mass which coats submerged surfaces. A slimy, rust-colored mass on the interior surface of flush tanks or water closets indicates the presence of iron bacteria.

Iron bacteria can reduce the carrying capacity of water pipes by increasing frictional losses. They may impart an unpleasant taste and odor to the water or discolor and spot fabrics, plumbing fixtures, and clog pumps. A detectable slime also builds up on any surface with which the water containing these organisms comes in contact. Iron bacteria may be concentrated in a specific location and may periodically break loose and appear at the faucet in detectable amounts of rust.

Iron-removal filters or water softeners can remove iron bacteria; however, they often become clogged and fouled because of the slime buildup. A disinfecting solution such as chlorine bleach should be injected into the water to control the growth of iron bacteria. Such a solution causes a chemical reaction which allows an iron precipitate to form. This precipitate can be removed with a suitable fine filter.

Softening

Water softening is a process for the removal of the minerals, primarily calcium and magnesium, which cause hardness.

Softening of hard water is desirable if

1. Large quantities of soap are needed to produce a lather.
2. Hard scale is formed on cooking utensils or laundry basins.
3. Hard, chalklike formations coat the interiors of piping or water tanks.
4. Heat-transfer efficiency through the walls of the heating element or exchange unit of the water tank is reduced.

The buildup of scale will cause an appreciable reduction in pipe capacities and pressures. The appearance of excessive scale from hard waters will also be esthetically objectionable. Experience has shown that hardness values greatly in excess of 200 mg/ℓ (12 grains per gallon) may cause some problems in the household.

Water may be softened by either the ion-exchange or the lime-soda ash process, but both processes increase the sodium content of the water and may make it unsuitable for people on a low-sodium diet.

Ion Exchange

The ion-exchange process causes a replacement of the calcium or magnesium ions by sodium ions. The process takes place when the hard water containing calcium or magnesium compounds comes in contact with an exchange medium. The materials used in the process of ion exchange are insoluble, granular materials that possess a unique property of ion exchange. Ion-exchange material may be classed as follows: glauconite (or greensand); precipitated synthetic, organic (carbonaceous), and synthetic resins; or gel zeolites. The last two are the most commonly used for domestic purposes.

The type of ion-exchange material used is determined by the type of water treatment required. For example, when a sodium zeolite is used to soften water by exchanging the sodium ion for calcium and magnesium ions in the hard water, the zeolite sodium ions eventually become of insufficient quantity to effect the exchange. After a certain period of time determined by the exchange rate, the exchange material must be regenerated. The sodium ion is restored to the zeolite by passing a salt (NaCl) or brine solution through the bed. The salt solution used must contain the same type of ions which were displaced by the calcium and magnesium. The solution causes a reversal of the ion-exchange process, restoring the exchange material to its original condition.

The type of regenerating material or solution which *must* be used depends upon the type of exchange material in the ion-exchange column.

The ion-exchange method of softening water is relatively simple and can be easily adapted to the small or individual water supply system. Only a portion of the hard water needs to be passed through the softening process because the exchange process produces water of zero hardness. The processed water can then be mixed with the hard water in proportions to produce a final water with a hardness between 50 to 80 mg/ℓ (3 to 5 grains per gallon). Waters with a turbidity of more than 10 Jackson units (an arbitrary measure of the light-scattering properties of suspended particles in water) should be properly treated for removal to increase the effectiveness and the efficiency of the softening process.

Ion-exchange softeners are commercially available for individual water systems. Their capacities range from about 85,000 to 550,000 milligrams of hardness that can be removed for each cubic foot of the ion-exchange material. Water softeners should be installed only by responsible persons in strict accordance with the instructions from the manufacturer and applicable codes. The materials and workmanship should be guaranteed for a specified period of time. First consideration in securing ion-exchange water-softening equipment should be given to those companies providing responsible servicing dealers permanently located within a reasonable distance from the water supply system. *Note.* Zeolite softening is not recommended if any of the water consumers, for medical reasons, are on a restricted sodium diet.

Lime-Soda Ash Process

The use of the lime-soda ash process or the addition of other chemicals is not practical for a small water supply system. Water used for laundry purposes, however, may be softened at the time of use by the addition of certain chemicals such as borax, washing soda, trisodium phosphate, or ammonia. Commercial softening or water conditioning compounds of unknown composition should *not* be used in water intended for drinking or cooking until the advice of the State or local health department is obtained regarding their safety.

Fluoridation

The presence of trace quantities of fluoride in the diet has been found beneficial in reducing dental caries in children and young adults. Water is currently an economical medium through which these trace quantities can be assimilated through body processes into the enamel of the teeth.

Equipment for fluoridating even the smallest home water supplies has been developed and used for several years. It is recommended, however, that the installer maintain the home fluoridator and test the treated water for fluoride level. It is an economical and reliable means of providing fluoridated water if the operation and maintenance of the fluoridating equipment are combined with other home water supply services; i.e., softening, iron removal, chlorination, and the like.

When a question of specific application arises, the recommendations of the State or local health department should be followed.

Tastes and Odors

Tastes and odors present in an individual water supply system fall into two general classes – natural and man made. Some natural causes may be traced to the presence of or contact of water with algae, leaves, grass, decaying vegetation, dissolved gases, and slime-forming organisms. Some of the manmade causes of taste and odor may be attributed to the presence of chemicals or sewage.

Water having a "rotten egg" odor indicates the presence of hydrogen sulfide and is commonly referred to as sulfur water. In addition to its objectionable odor, sulfur water may cause a black stain on plumbing fixtures. Hydrogen sulfide is very corrosive to common metals and will react with iron, copper, or silver to form the sulfides of these metals.

Depending upon the cause, taste and odor can be removed or reduced by aeration or by treatment with activated carbon, copper sulfate, or an oxidizing agent such as chlorine.

Aeration is exposure of as much water surface as possible to the air. It is described in the section entitled "Aeration." Hydrogen sulfide can be removed by aeration or by a combination oxidization-filtration process. A simple iron-removal filter will also do a good job of removing this objectionable compound when small amounts are involved.

The activated carbon treatment consists of passing the water to be treated through granular carbon, or adding powdered activated carbon to the water. Activated carbon adsorbs (attracts to itself) large quantities of dissolved gases, soluble organics, and finely divided solids. It is therefore extremely effective in taste and odor control. Activated carbon can be used in carbon filters commercially available from the manufacturers or producers of water-conditioning or treatment equipment. The recommendations included with the filter should be followed.

Copper Sulfate. The most frequent source of taste and odors in an individual water supply system is algae. These minute plants produce certain biological byproducts which cause tastes and odors in the water. These tastes and odors may be accentuated when chlorine is added to the water. When they are present in a water supply their growth can be controlled by adding copper sulfate to the water source, as described in the section dealing with "Algae Control."

Because algae and other chlorophyll-containing plants need sunlight to grow, the storage of water in covered reservoirs inhibits their growth.

Chlorine. Chlorine is an effective agent in reducing tastes and odors present in water. The process used for the reduction of tastes and odors is the same as described in the section dealing with "Superchlorination-Dechlorination."

Corrosion Control

The control of corrosion is important not only to continuous and efficient operation of the individual water system but also to delivery of properly conditioned water that has not picked up trace quantities of metals that may be hazardous to health. Whenever corrosion is minimized there is an appreciable reduction in the maintenance and possible replacement of water pipes, water heaters, or other metallic appurtenances of the system.

Corrosion is commonly defined as an electrochemical reaction in which metal deteriorates or is destroyed when in contact with elements of its environment such as air, water, or soil. Whenever this reaction occurs there is a flow of electric current from the corroding portion of the metal toward the electrolyte or conductor of electricity, such as water or soil. The point at which current flows from the metal into the electrolyte is called the "anode" and the point at which current flows away from the electrolyte is called the "cathode." Any characteristic of the water which tends to allow or increase the rate of this electrical current will increase the rate of corrosion. The important characteristics of a water that affect its corrosiveness include the following:

1. *Acidity.* A measure of the water's ability to neutralize alkaline materials. Water with acidity or low alkalinity (a measure of the concentration of alkaline materials) tends to be corrosive.
2. *Conductivity.* A measure of the amount of dissolved mineral salts. An increase in conductivity promotes flow of electrical current and increases the rate of corrosion.
3. *Oxygen content.* Amount dissolved in water promotes corrosion by destroying the thin protective hydrogen film that is present on the surface of metals immersed in water.
4. *Carbon dioxide.* Forms carbonic acid, which tends to attack metallic surfaces.
5. *Water temperatures.* The corrosion rate increases with water temperature.

Corrosion and Scale Relationship

Corrosion and scale are associated problems, but their effect and cause should not be confused. The essential effect of corrosion is to destroy metal; scale, on the other hand, tends to clog open sections and line surfaces with deposits. The products of corrosion often contribute to scale formation and aggravate the problem of its treatment.

Prevention of Corrosion

When corrosion is caused by the acidity of the water supply, it can be effectively controlled by installing an acid neutralizer ahead of a water softener. Another method of controlling corrosion is that of feeding small amounts of commercially available film-forming materials such as polyphosphates or silicates. Other methods for controlling corrosion are the installation of dielectric or insulating unions, reduction of velocities and pressures, removal of oxygen or acid constituents, chemical treatment to decrease the acidity, or the use of nonmetallic piping and equipment.

pH Correction or Neutralizing Solution

The pH of water may be increased by feeding a neutralizing solution so that it no longer attacks parts of the water system or contributes to electrolytic corrosion. Neutralizing solutions may be prepared by mixing soda ash (58% light grade) with water – 3 pounds soda ash to 4 gallons of water. This solution may be fed into the water supply with feeders as described under "Chlorination," and may be mixed with chlorine solutions to accomplish both pH correction and disinfection with the same equipment. Soda ash is available at chemical supply houses.

Nuisance Organisms

Organisms that have been known to cause problems in water supplies include several species of algae, protozoa, and diatoms that produce tastes and odors and clog filters. Iron bacteria plug water well intakes and clog pipes in distribution systems (see Iron Bacteria, p. 84). Still other nuisance organisms are copepods, whose eggs pass through filters; midge larvae or bloodworms; and snails and mollusca. These organisms vary in complexity and size. They are uncommon or absent in ground water, but are common in surface waters.

Perhaps none of the organisms is injurious to health. Interference with water treatment processes, and unpleasant taste, odor and appearance constitute the chief complaints against them.

Control of Algae

Growths of algae can be controlled by treating the water with copper sulfate (blue stone or blue vitriol) or, when feasible, by covering the storage unit to exclude sunlight. Maintenance of an adequate chlorine residual will effectively control the growth of algae and other organisms wherever storage is covered and protected from contamination. The particular control method, or combination of methods, is determined by studying each case to assess the probability for success and the costs involved.

Copper sulfate has been used successfully for the control of algae since 1900. Temperature, pH and alkalinity all affect the solubility of copper in water. From this it can be seen that the dosage required depends on the chemistry of the water treated and the susceptibility to copper of the particular nuisance organism present. Dosage rates of 1 ounce of copper sulfate ($CuSO_4 \cdot 5H_2O$) for each 25,000 gallons of water have proven effective where the total alkalinity of the water does not exceed 40 mg/l (40 ppm). For more alkaline waters, the dosage can be increased to 5.5 pounds of copper sulfate per acre of surface water treated regardless of depth.

Frequency of treatment depends on temperature, amount of sunlight, and nutrients in the water. Systematic application of the calculated amount of chemical over the entire surface area ensures that serious algal blooms do not reappear. Several treatments per season are generally required, with treatments as frequent as twice a month during the growing season not being unusual.

The most practical method of application for small ponds is by spraying a solution on the surface. Or, a burlap bag of copper sulfate can be dragged through the water. Rapid and uniform distribution of the chemical is important.

It should be noted that sudden kill of heavy growths of algae may be followed by decomposition on a scale that depletes the oxygen content of the water. If the removal of oxygen is excessive, a fish kill may result.

Any chemical applied to control a problem with nuisance organisms must be used with caution. The concentrations recommended above will affect only a portion of the biotic system. Excessive amounts of chemical may endanger other life systems in the environment. If there is any doubt about the effects which treatment might have on other life systems, advice should be sought from responsible environmental agencies.

Aeration

Aeration is the process of bringing about the intimate contact between air and a liquid such as water.

Many methods are available for obtaining effective aeration, including spraying water into the air, allowing water to fall over a spillway in a turbulent stream, or distributing water in multiple streams or droplets through a series of perforated plates. Although the aeration of water may be accomplished in an open system, adequate precautions should be exercised to eliminate possible external contamination of the water. Whenever possible, a totally enclosed system should be provided.

Aeration may be used to oxidize iron or manganese and remove odors from water, such as those caused by hydrogen sulfide and algae. It is also effective in increasing the oxygen content of water deficient in dissolved oxygen. The flat taste of cistern water and distilled water may be improved by adding oxygen. Carbon dioxide and other gases that increase the corrosiveness of water can be eliminated largely by effective aeration, although the increase in corrosion because of increased oxygen may partially offset the advantage of the decrease in carbon dioxide.

Aeration of water results in partial oxidation of its dissolved iron or manganese and thereby changes the iron into an insoluble form. Sometimes a short period of storage permits the insoluble material to settle; at other times the precipitated iron or manganese cannot be removed successfully except by filtration.

A simple cascade device or a coke tray (wire-bottom trays filled with activated carbon) aerator can be incorporated into a water supply system. In addition to aerating, the coke tray will reduce tastes and odors.

Insects such as the chironomus fly may lay eggs in the stagnant portion of the aerator tray. The eggs develop into small red worms, which is the larvae stage of this insect. Proper encasement of the aerator prevents the development of this situation. Adequate screening will provide, in addition, protection from windblown debris.

Part V

Pumping, Distribution, and Storage

PUMPING

Types of Well Pumps

Three types of pumps are commonly used in individual water distribution systems. They are the positive displacement, the centrifugal, and the jet. These pumps can be used in a water system utilizing either a ground or surface source. It is desirable in areas where electricity or other power (gasoline, diesel oil, or windmill) is available to use a power-operated pump. When a power supply is not available, a hand pump or some other manual method of supplying water must be used.

Special types of pumps with limited application for individual water-supply systems include air lift pumps and hydraulic rams.

Positive Displacement Pumps

The positive displacement pump forces or displaces the water through a pumping mechanism. These pumps are of several types.

One type of positive displacement pump is the reciprocating pump. This pump consists of a mechanical device which moves a plunger back and forth in a closely fitted cylinder. The plunger is driven by the power source, and the power motion is converted from a rotating action to a reciprocating motion by the combined work of a speed reducer, crank, and a connecting rod. The cylinder, composed of a cylinder wall, plunger, and check valve, should be located near or below the static water level to eliminate the need for priming. The pumping action begins when the water enters the cylinder through a check valve. When the piston moves, the check valve closes, and in so doing forces the water through a check valve in the plunger. With each subsequent stroke, the water is forced toward the surface through the discharge pipe.

Another type of positive displacement pump is the helical or spiral rotor. The helical rotor consists of a shaft with a helical (spiral) surface which rotates in a rubber sleeve. As the shaft turns, it pockets or traps the water between the shaft and the sleeve and forces it to the upper end of the sleeve.

Other types of positive displacement pumps include the regenerative turbine type. It incorporates a rotating wheel or impeller which has a series of blades or fins (sometimes called buckets) on its outer edge and a stationary enclosure called a raceway or casting. Pressures several times that of pumps relying solely on centrifugal force can be developed.

Centrifugal Pumps

Centrifugal pumps are pumps containing a rotating impeller mounted on a shaft turned by the power source. The rotating impeller increases the velocity of the water and discharges it into a surrounding casing shaped to slow down the flow of the water and convert the velocity to pressure. This decrease of the flow further increases the pressure.

Each impeller and matching casing is called a stage. The number of stages necessary for a particular installation will be determined by the pressure needed for the operation of the water system, and the height the water must be raised from the surface of the water source.

When the pressure is more than can be practicably or economically furnished by a single stage, additional stages are used. A pump with more than one stage is called a multistage pump. In a multistage pump water passes through each stage in succession, with an increase in pressure at each stage.

Multistage pumps commonly used in individual water systems are of the turbine and submersible types.

Turbine Pumps   The vertical-drive turbine pump consists of one or more stages with the pumping unit located below the drawdown level of the water source. A vertical shaft connects the pumping assembly to a drive mechanism located above the pumping assembly. The discharge casing, pumphousing, and inlet screen are suspended from the pump base at the ground surface. The weight of the rotating portion of the pump is usually suspended by a thrust bearing located in the pump head. The intermediate pump bearings may be lubricated by either oil or water. From a sanitary point of view, lubrication of pump bearings by water is preferable, since lubricating oil may leak and contaminate the water.

Submersible Pumps.   When a centrifugal pump is driven by a closely coupled electric motor constructed for submerged operation as a single unit, it is called a submersible pump. (See fig. 15.) The electrical wiring to the submersible motor must be waterproof. The electrical control should be properly grounded to minimize the possibility of shorting and thus damaging the entire unit. The pump and motor assembly are supported by the discharge pipe; therefore, the pipe should be of such size that there is no possibility of breakage.

The turbine or submersible pump forces water directly into the distribution system; therefore, the pump assembly must be located below the maximum drawdown level. This type of pump can deliver

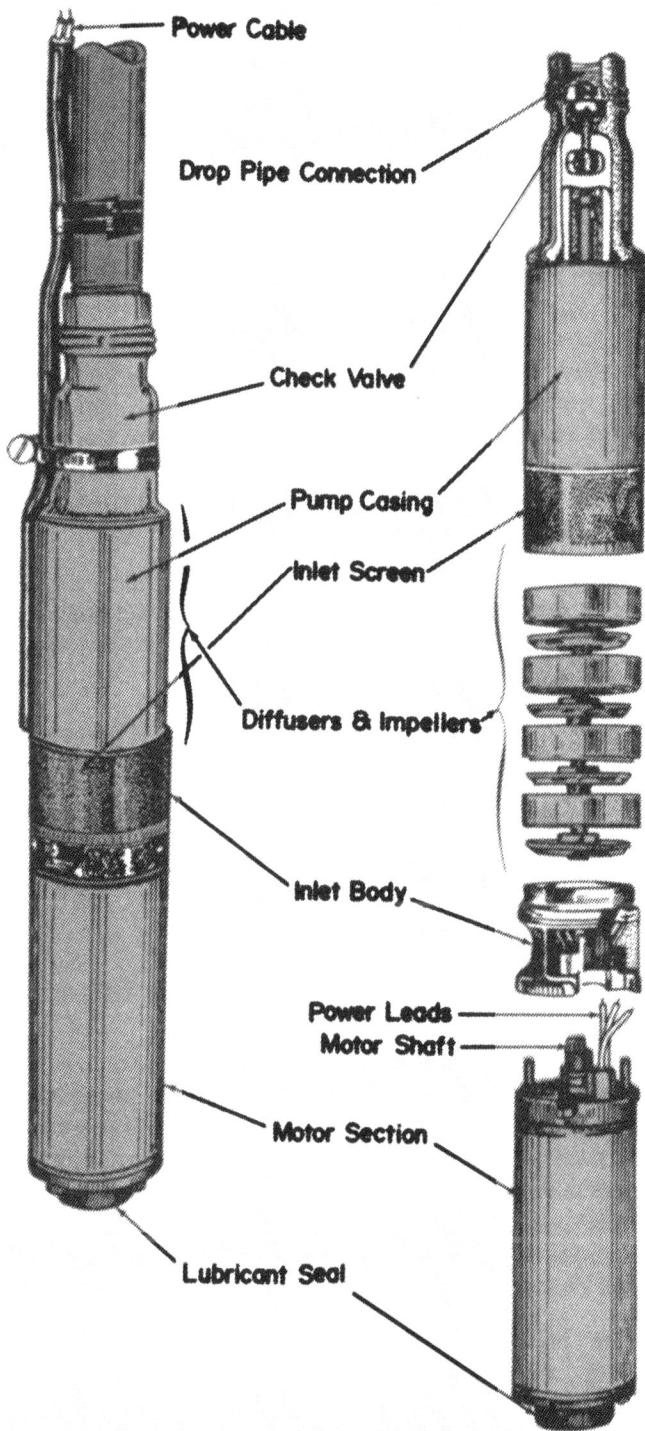

Power Cable

Drop Pipe Connection

Check Valve

Pump Casing

Inlet Screen

Diffusers & Impellers

Inlet Body

Power Leads

Motor Shaft

Motor Section

Lubricant Seal

FIGURE 15.   Exploded view of submersible pump.

water across a wide range of pressures with the only limiting factor being the size of the unit and the horsepower applied. When sand is present or anticipated in the water source, special precautions should be taken before this type of pump is used since the abrasion action of the sand during pumping will shorten the life of the pump.

Jet (Ejector) Pumps

Jet pumps are actually combined centrifugal and ejector pumps. A portion of the discharged water from the centrifugal pump is diverted through a nozzle and venturi tube. A pressure zone lower than that of the surrounding area exists in the venturi tube; therefore, water from the source (well) flows into this area of reduced pressure. The velocity of the water from the nozzle pushes it through the pipe toward the surface where the centrifugal pump can lift it by suction. The centrifugal pump then forces it into the distribution system. (See fig. 16.)

Selection of Pumping Equipment

The type of pump selected for a particular installation should be determined on the basis of the following fundamental considerations.

1. Yield of the well or water source.
2. Daily needs and instantaneous demand of the users.
3. The "usable water" in the pressure or storage tank.
4. Size and alinement of the well casing.
5. Total operating head pressure of the pump at normal delivery rates, including lift and all friction losses.
6. Difference in elevation between ground level and water level in the well during pumping.
7. Availability of power.
8. Ease of maintenance and availability of replacement parts.
9. First cost and economy of operation.
10. Reliability of pumping equipment.

The rate of water delivery required depends on the time of effective pump operation as related to the total water consumption between periods of pumping. Total water use can be determined from table 1, page 15. The period of pump operation depends upon the quantity of water on hand to meet peak demands and the storage available. If the well yield permits, a pump capable of meeting the peak demand (see table 9, page 124) should be used.

When the well yield is low in comparison to peak demand requirements, an appropriate increase in the storage capacity is required. The life of an electric drive motor will be reduced when there is. excessive starting and stopping. The water system, therefore, should be designed so that the interval between starting

and stopping is as long as is practicable but not less than 1 minute.

Counting the number of fixtures in the home permits a ready determination of required pump capacity from figure 17. For example, a home with kitchen sink, water closet, bathtub, wash basin, automatic clothes washer, laundry tub and two outside hose bibs, has a total of eight fixtures. Referring to the figure, it is seen that eight fixtures correspond to a recommended pump capacity between 9 and 11 gallons per minute. The lower value should be the minimum. The higher value might be preferred if additional fire protection (see p. 17) is desired, or if garden irrigation (see "Lawn Sprinkling," p. 16) or farm use (table 7) is contemplated.

(*Note*: This simple calculation does not take into account the possibility that low well capacity may limit the size of pump that should be installed. In this case, the system can be reinforced by providing additional storage to help cover periods of peak demand. See "Storage," p. 124.)

The total operating head of a pump consists of the lift (vertical distance from pumping level of the water source to the pump), the friction losses in the pipe and fittings from water source to pump, and the discharge pressure at the pump. (See fig. 18.)

Pumps that cannot be wholly submerged during pumping are dependent on suction to raise the water from the source by reducing the pressure in the pump column, or creating a suction. The vertical distance from the source (pumping level) to the axis of the pump is called the suction lift, and for practical purposes cannot exceed between 15 and 25 feet, depending on the design of the pump and the altitude above sea level where it is used.

Shallow well pumps should be installed with a foot valve at the bottom of the suction line or with a check valve in the suction line in order to maintain pump prime.

The selection of a pump for any specific installation should be based on *competent advice.* Authorized factory representatives of pump manufacturers are among those best qualified to provide this service.

Sanitary Protection of Pumping Facilities

The pump equipment for either power-driven or manual systems should be so constructed and installed as to prevent the entrance of contamination or objectionable material either into the well or into the water that is being pumped. The following factors should be considered.

1. Designing the pump head or enclosure so as to prevent pollution of the water by lubricants or other maintenance materials used during operation of the equipment. Pollution from hand contact, dust, rain, birds, flies, rodents or animals, and similar sources should be

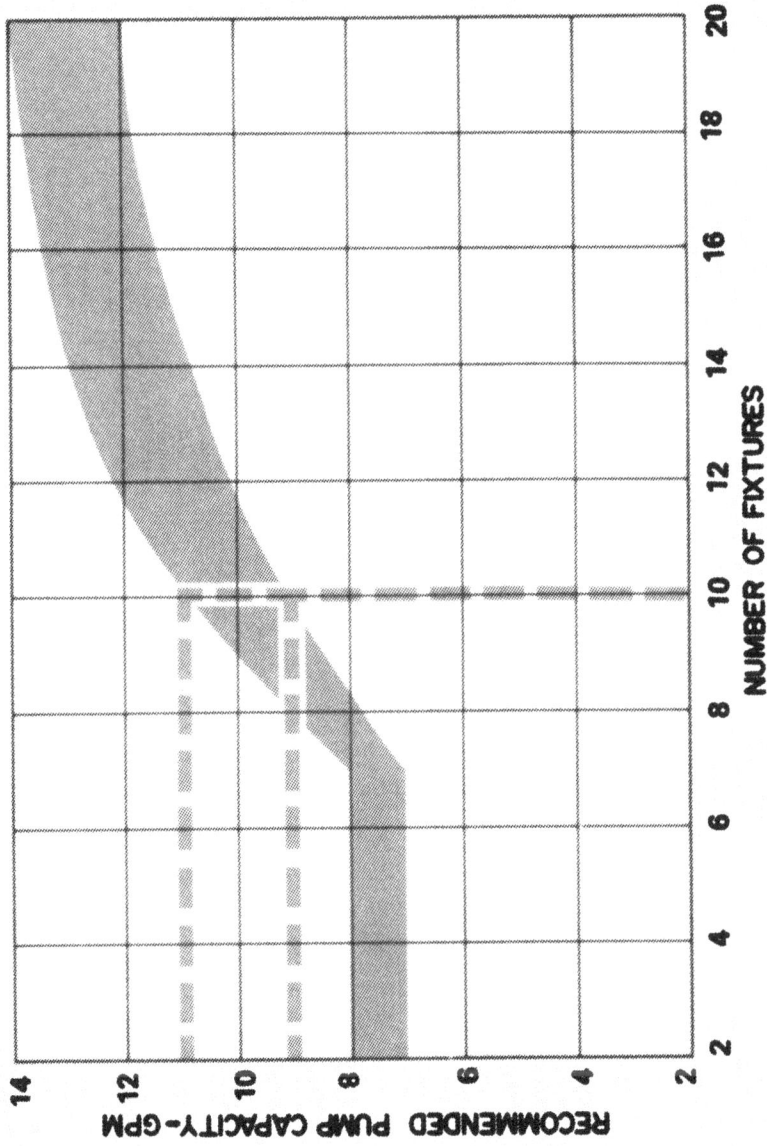

FIGURE 17. Determining recommended pump capacity.

TABLE 7. — *Information on pumps*

Type of pump	Practical suction lift[1]	Usual well-pumping depth	Usual pressure heads	Advantages	Disadvantages	Remarks
Reciprocating						
1. Shallow well	22-25 ft	22-25 ft	100-200 ft	1 Positive action	1 Pulsating discharge	1. Best suited for capacities of 5-25 gpm against moderate to high heads.
2. Deep well	22-25 ft	Up to 600 ft	Up to 600 ft. above cylinder	2 Discharge against variable heads	2 Subject to vibration and noise	2 Adaptable to hand operation
				3 Pumps water containing sand and silt	3 Maintenance cost may be high	3 Can be installed in very small diameter wells (2" casing).
				4 Especially adapted to low capacity and high lifts	4 May cause destructive pressure if operated against closed valve	4 Pump must be set directly over well (deep well only)
Centrifugal						
1 Shallow well						
a Straight centrifugal (single stage)	20 ft max	10-20 ft	100-150 ft	1 Smooth, even flow	1 Loses prime easily	1 Very efficient pump for capacities above 60 gpm and heads up to about 150 ft
				2 Pumps water containing sand and silt	2 Efficiency depends on operating under design heads and speed	
				3 Pressure on system is even and free from shock		
				4 Low-starting torque		
				5 Usually reliable and good service life		
b Regenerative vane turbine type (single impeller)	28 ft max	28 ft	100-200 ft	1 Same as straight centrifugal except not suitable for pumping water containing sand or silt	1 Same as straight centrifugal except maintains priming easily	1 Reduction in pressure with increased capacity not as severe as straight centrifugal
				2 They are self-priming.		
2 Deep well						
a Vertical line shaft turbine (multi-stage)	Impellers submerged.	50-300 ft	100-800 ft	1 Same as shallow well turbine	1 Efficiency depends on operating under design head and speed	
				2 All electrical components are accessible, above ground	2 Requires straight well large enough for turbine bowls and housing	
					3 Lubrication and alignment of shaft critical	
					4 Abrasion from sand	

				Advantages	Disadvantages	Remarks
b Submersible turbine (multistage)	Pump and motor submerged	50-400 ft	50-400 ft	1 Same as shallow well turbine 2 Easy to frost-proof installation 3 Short pump shaft to motor 4 Quiet operation 5 Well straightness not critical	1 Repair to motor or pump requires pulling from well 2 Sealing of electrical equipment from water vapor critical 3 Abrasion from sand	1 3500 **RPM** models, while popular because of smaller diameters or greater capacities, are more vulnerable to wear and failure from sand and other causes
Jet 1 Shallow well	15-20 ft below ejector	Up to 15-20 ft below ejector	80-150 ft	1 High capacity at low heads 2 Simple in operation 3 Does not have to be installed over the well 4 No moving parts in the well	1 Capacity reduces as lift increases 2 Air in suction or return line will stop pumping	
2 Deep well	15-20 ft below ejector	25-120 ft 200 ft max	80-150 ft	1 Same as shallow well jet 2 Well straightness not critical	1 Same as shallow well jet 2 Lower efficiency, especially at greater lifts	1 The amount of water returned to ejector increases with increased lift – 50% of total water pumped at 50-ft lift and 75% at 100-ft lift
Rotary 1 Shallow well (gear type)	22 ft	22 ft	50-250 ft	1 Positive action 2 Discharge constant under variable heads 3 Efficient operation	1 Subject to rapid wear if water contains sand or silt 2 Wear of gears reduces efficiency	
2 Deep well (helical rotary type)	Usually submerged	50-500 ft	100-500 ft	1 Same as shallow well rotary 2 Only one moving pump device in well	1 Same as shallow well rotary except no gear wear	1 A cutless rubber stator increases life of pump flexible drive coupling has been weak point in pump Best adapted for low capacity and high heads

[1] Practical suction lift at sea level Reduce lift 1 foot for each 1,000 ft above sea level

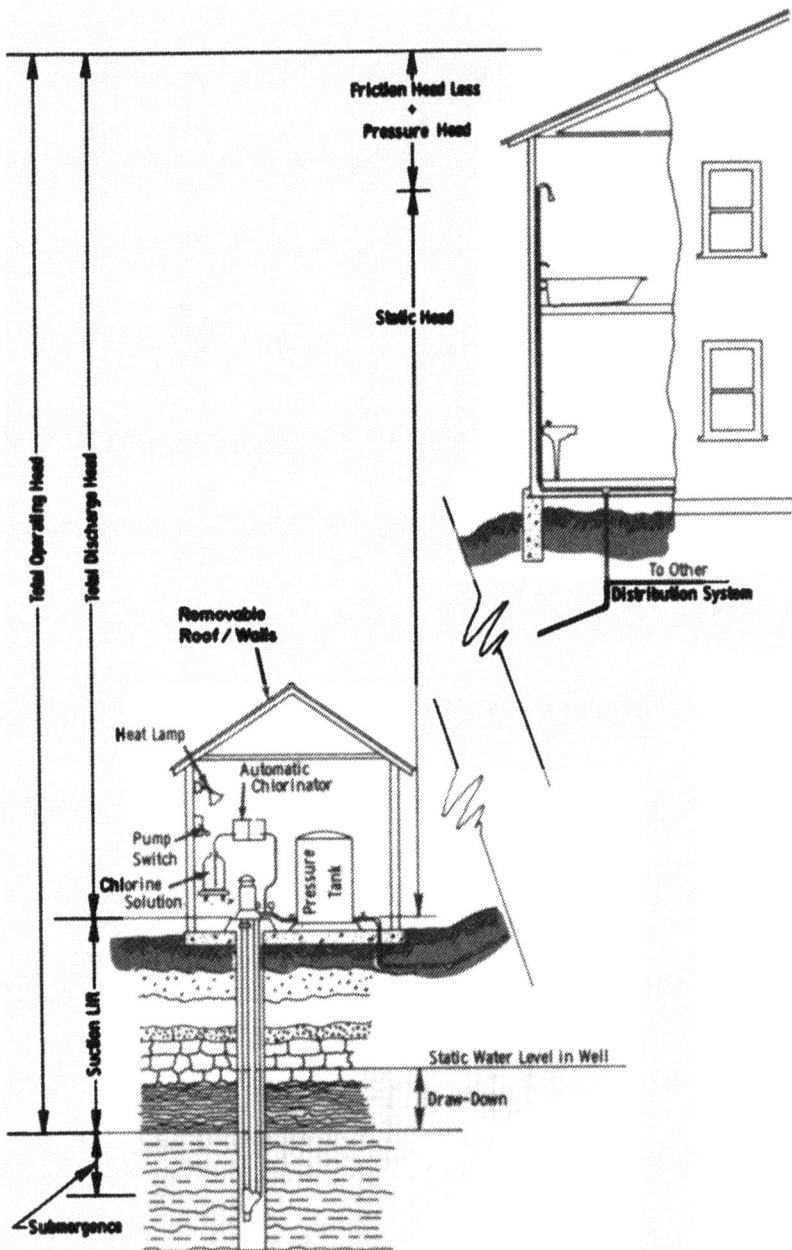

FIGURE 18. Components of total operating head in well pump installations.

102

prevented from reaching the water chamber of the pump or the source of supply.

2. Designing the pump base or enclosure so as to facilitate the installation of a sanitary well seal within the well cover or casing.

3. Installation of the pumping portion of the assembly near or below the static water level in the well so that priming will not be necessary.

4. Designing for frost protection, including pump drainage within the well when necessary.

5. Overall design consideration so as to best facilitate necessary maintenance and repair, including overhead clearance for removing the drop pipe and other accessories.

When planning for sanitary protection of a pump, specific types of installations must be considered. The following points should be considered for the different types of installations.

Check Valves. The *only* check valve between the pump and storage should be located within the well (see fig. 24, p. 114), or at least upstream from any portion of a buried discharge line. This will assure that the discharge line at any point where it is in contact with soil or a potentially contaminated medium will remain under positive system pressure – whether or not the pump is operating. There should be *no* check valve at the inlet to the pressure tank or elevated storage tank. This requirement would not apply to a concentric piping system, with the external pipe constantly under system pressure. (See fig. 14, p. 69; fig. 22, p. 112; and fig. 23, p. 113.)

Many pumps (submersibles, jets) normally have check valves installed within the well.

Well Vents. A well vent is recommended on all wells not having a packer-type jet pump. The vent prevents a partial vacuum inside the well casing as the pump lowers the water level in the well. (The packer-type jet installation cannot have a well vent, since the casing is subjected to positive system pressure.) The well vent – whether built into the sanitary well cover or conducted to a point remote from the well – should be protected from mechanical damage, have watertight connections, and be resistant to corrosion, vermin, and rodents. (See fig. 7, p. 40; fig. 9, p. 47; and fig. 24, p. 114.)

The opening of the well vent should be located not less than 24 inches above the highest known flood level. It should be screened with durable and corrosion-resistant materials (bronze or stainless steel no. 24 mesh) or otherwise constructed so that openings exclude insects and vermin.

Miscellaneous Certain types of power pumps require that the water be introduced into the pumping system, either to prime the pump or to lubricate rubber bearings that have become dry while the pump was inoperative. Water used for priming or lubricating should be free of contamination.

It is desirable to provide a water-sampling tap on the discharge line from power pumps.

Installation of Pumping Equipment

Where and how the pump and power unit are mounted depend primarily on the type of pump employed. The vertical turbine centrifugal pump, with power source located directly over the well and the pumping assembly submerged within the well, is gradually being replaced by the submersible unit, where both the power unit (electric motor) and the pump are submerged within the well. Similarly, the jet pump is gradually giving way to the submersible pump — especially for deeper installations — because of the latter's inherently superior performance and better operating economy.

Vertical Turbine Pumps. In the vertical turbine pump installation, the power unit (usually an electric motor) is installed directly over the well casing. The pump portion is submerged within the well, and the two are connected by a shaft enclosed within the pump column. The pump column supports the bearing system for the drive shaft and conducts the pumped water to the surface. (See fig. 19.)

Since the long shaft must rotate at high speed (1,800 to 3,600 rpm), correct alinement of the motor, shaft, and pump is vital to good performance and long life of the equipment. There are two main points to consider in obtaining a proper installation:

1. Correct and stable positioning of the power unit.
2. Verticality and straightness of the pump column within the well.

Since concrete slabs tend to deteriorate, settle, or crack from weight and vibration, it is usually better to attach the discharge head to the well casing. Figure 19 shows one way to accomplish this. For smoothest operation and minimum wear, the plate (and discharge head) should be mounted perpendicular to the axis of the pump column as pump and column hang in the well. If the casing is perfectly plumb, the pump column axis and the well axis coincide, and a perfect installation results. It sometimes happens, though, that the well is not plumb, or that it is crooked. In this case, it is necessary to adjust the position of the plate so that the axis of the pump column lies as close as possible to the axis of the well. If there is enough room inside the casing (and this is one of the reasons for installing larger casing), there is a better chance that pump and column will be able to hang plumb -- or at least be able to operate smoothly. Once the correct position of the plate is determined, it is welded to the well casing. The discharge head is then bolted securely to the support plate.

As explained under "Sanitary Construction of Wells" on p. 48,

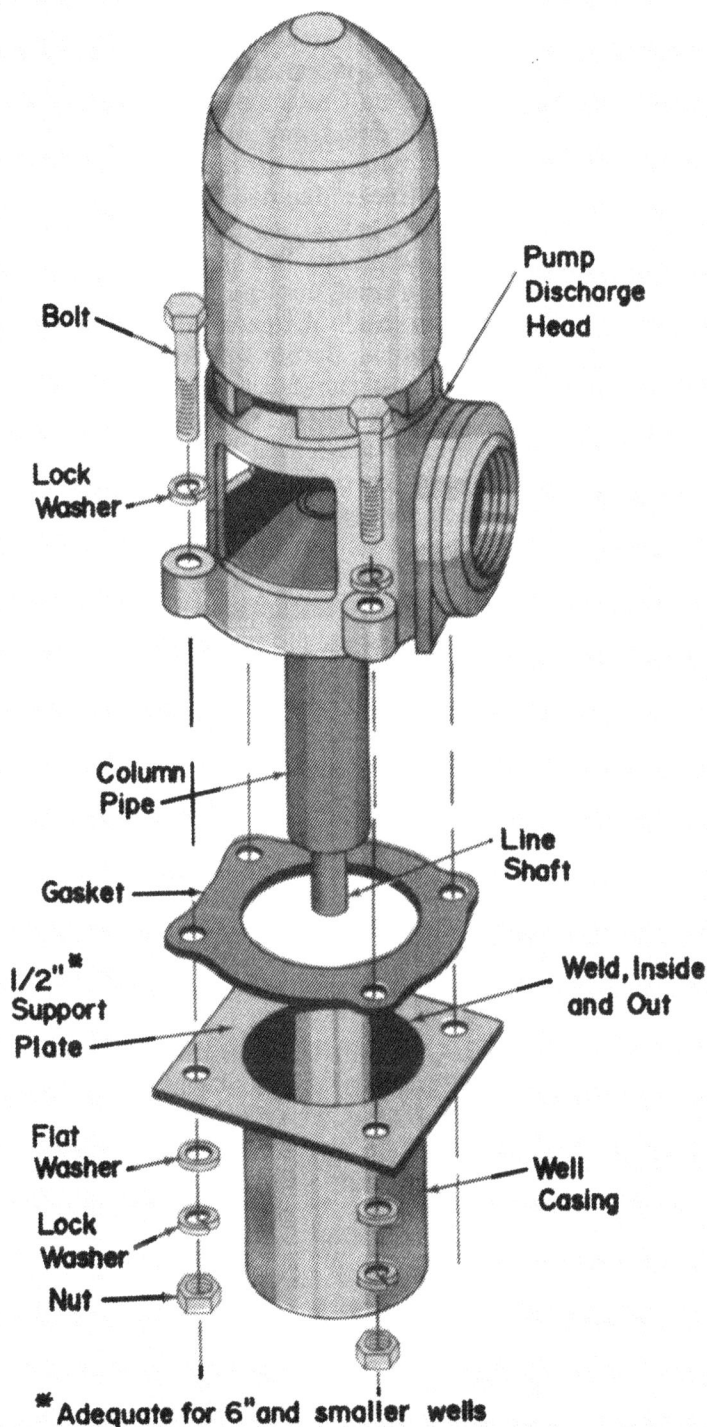

Bolt

Lock
Washer

Column
Pipe

Gasket

1/2"*
Support
Plate

Flat
Washer

Lock
Washer

Nut

Pump
Discharge
Head

Line
Shaft

Weld, Inside
and Out

Well
Casing

*Adequate for 6"and smaller wells

FIGURE 19. Vertical (line shaft) turbine pump mounted on well casing.

sanitary well seals or covers are available for installation to seal the well casing against contamination entering at this point. Some designs, however, make it difficult or impossible to measure water levels within the well. This deficiency should be corrected by welding to the side of the casing an access pipe, which permits introduction of a water-level measuring device. A hole is first cut in the casing at a point far enough below the top to permit clear access past the discharge head of the pump. The angle between the access pipe and the casing should be small enough to permit free entry of the measuring line. Minimum inside diameter of the pipe should be 3/4 inch, and larger when possible. Before welding the pipe in place, any sharp edges around the hole through the casing should be filed smooth so that the measuring device will slide freely through without catching or becoming scratched. An inclination angle of one unit horizontal to four units vertical provides a good access — or, in other words, for each foot down from the top of the casing, the access pipe will be inclined outward 3 inches horizontally from the top.

Some engineers and well service technicians recommend that all wells be equipped with such access pipes because of the ease of introducing and withdrawing measuring devices and because the pipe permits chemical treatment of the well without removing the sanitary well seal and pump.

The welding around the access pipe should be at least as thick and resistant to corrosion as the well casing itself. This is especially important if the connection will be located below the ground surface.

Submersible Pumps. Because all moving parts of the submersible pump are located within the well in a unit, this pump can perform well in casings that might be too crooked for vertical turbine pumps. If there is little difference between the inside casing diameter and the outside diameter of the pump, the pump might stick in the well casing, with the possibility that it could be damaged during installation. If there is any doubt about whether there is room, a "dummy" piece of pipe whose dimensions are slightly greater than those of the pump should first be run through the casing to make sure that the pump will pass freely to the desired depth of setting.

The entire weight of the pump, cable, drop pipe, column of water within the pipe, and reaction load when pumping must be supported by the drop pipe itself. It is important, therefore, that the drop pipe and couplings be of good quality steel, galvanized, and of standard weight. (See "Casing and Pipe," p. 42.) Cast-iron fittings should not be used where they must support pumps and pump columns.

The entire load of submersible pumping equipment is normally suspended from the sanitary well seal or cover. An exception to this would be the "pitless" installation. (See p. 109.)

Jet Pumps. Jet pumps may be installed directly over the well, or alongside it. Since there are no moving parts in the well, straightness and plumbness do not affect the jet pump's performance. The weight of equipment in the well is relatively light, being mostly pipe (often plastic), so that loads are supported easily by the sanitary well seal. There are also a number of good "pitless adapter" and "pitless unit" designs for both single and double pipe jet systems. (See p. 109.)

Hand Pumps. The pump heads on most force pumps are designed with a stuffing box surrounding the pump rod. This design provides reasonable protection against contamination. Ordinary lift pumps with slotted pump head tops are open to contamination and should not be used. The pump spout should be closed and directed downward.

The pump base should be designed to serve a twofold purpose: first, to provide a means of supporting the pump on the well cover or casing top; and second, to protect the well opening or casing top from the entrance of contaminated water or other harmful or objectionable material. The base should be of the solid, one-piece, recessed type, cast integrally with or threaded to the pump column or stand. It should be of sufficient diameter and depth to permit a 6-inch well casing to extend at least 1 inch above the surface upon which the pump base is to rest. The use of a flanged sleeve embedded in the concrete well cover or a flange threaded or clamped on the top of the casing to form a support for the pump base is recommended. Suitable gaskets should be used to insure tight closure.

The protective closing of the pump head, together with the pollution hazard incident to pump priming, makes it essential that the pump cylinders be so installed that priming will not be necessary.

Pumphousing and Appurtenances

A pumphouse installed above the surface of the ground should be used. (See fig. 20.) The pumproom floor should be of watertight construction, preferably concrete, and should slope uniformly away in all directions from the well casing or pipesleeve. It should be unnecessary to use an underground discharge connection if an insulated, heated pumphouse is provided. For individual installations in rural areas, two 60-watt light bulbs, a thermostatically controlled electric heater, or a heating cable will generally provide adequate protection when the pumphouse is properly insulated.

In areas where power failures may occur, an emergency,

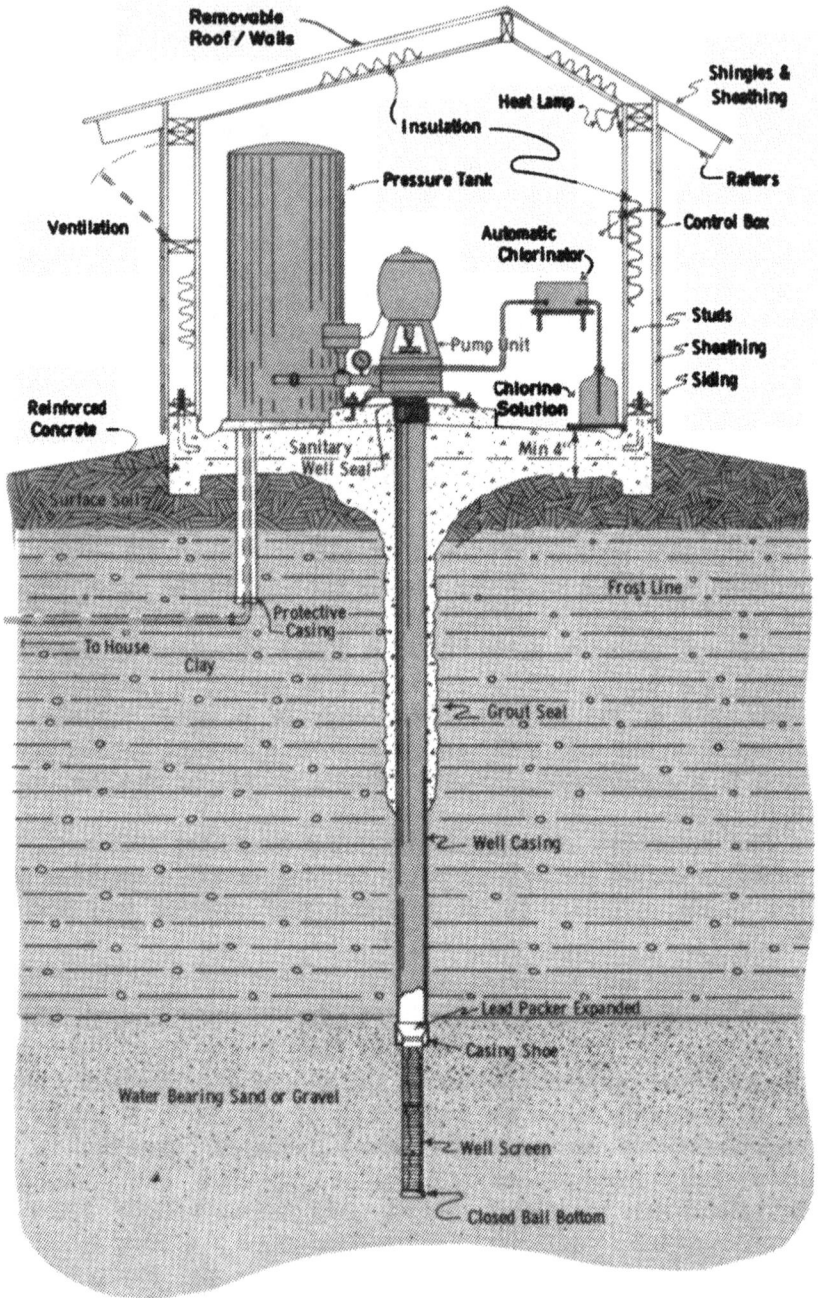

FIGURE 20.   Pumphouse.

gasoline-driven power supply or pump should be considered. A natural disaster, such as a severe storm, hurricane, tornado, blizzard, or flood, may cut off power for hours or even days. A gasoline, power-driven electrical unit could supply the power requirements of the pump, basic lighting, refrigeration, and other emergency needs.

Lightning Protection

Voltage and current surges produced in powerlines by nearby lightning discharges constitute a serious threat to electric motors. The high voltage can easily perforate and burn the insulation between motor windings and motor frame. The submersible pump motor is somewhat more vulnerable to this kind of damage because it is submerged in ground water – the natural "ground" sought by the lightning discharge. Actual failure of the motor may be immediate, or it may be delayed for weeks or months.

There are simple lightning arresters available to protect motors and appliances from "near miss" lightning strikes. (They are seldom effective against direct hits.) The two types available are the valve type and the expulsion type. The valve type should be preferred because its "sparkover" voltage remains constant with repeated operation.

Just as important as selecting a good arrester is installing it properly. The device must be installed according to instructions from the manufacturer and connected to a *good* ground. In the case of submersible pumps, this good ground can be achieved by connecting the ground terminal of the arrester to the submersible pump motor frame by means of a no. 12 stranded bare copper wire. The low resistance (1 ohm or less) reduces the voltage surge reaching the motor windings to levels that it can resist.

If steel well casing extends *into the ground water,* the ground can be improved still further by also connecting the bare copper wire to the well casing. *IMPORTANT NOTE.* Connecting the ground terminal of the arrester to a copper rod driven into the ground does *not* satisfy grounding requirements. Similarly, if a steel casing that does not reach the ground water is relied upon, the arrester may be rendered ineffective.)

Additional advice on the location and installation of lightning arresters can be obtained from the power company serving the area.

Pitless Units and Adapters

Because of the pollution hazards involved, a well pit to house the pumping equipment or to permit accessibility to the top of the well is not recommended. Some States prohibit its use.

A commercial unit known as the "pitless adapter" is available to eliminate well pit construction. A specially designed connection between the underground horizontal discharge pipe and the vertical casing pipe makes it possible to terminate the permanent, watertight casing of the well at a safe he'ght (8 inches or more) above the final grade level. The underground section of the discharge pipe is permanently installed and it is not necessary to disturb it when repairing the pump or cleaning the well. (See figs. 21-24.)

There are numerous makes and models of pitless adapters and units available. Not all are of good design, and a few are not acceptable to some States. The State or local health department should be consulted first to learn what is acceptable.

Both the National Sanitation Foundation[1] and the Water Systems Council[2] have adopted criteria intended to assure that quality materials and workmanship are employed in the manufacture and installation of these devices. Unfortunately, the safety of these installations is highly dependent on the quality of workmanship applied during their attachment in the field. For this reason, additional precautions and suggestions are offered here.

There are two general types of pitless installations. One, the "pitless adapter," requires cutting a hole in the side of the casing at a predetermined depth below the ground surface (usually below the frost line). Into this opening there is inserted and attached a fitting to accommodate the discharge line from the pump. Its design varies according to whether it is for a pressure line alone or for both pressure and suction lines (two-pipe jet pump system with pump mounted away from well). The other part of the adapter, mounted inside the well, supports the pumping components that are suspended in the well. Watertight connection is accomplished by a system of rubber seals compressed by clamps or by the weight of the equipment itself.

The second type – the "pitless unit" – requires cutting off the well casing at the required depth and mounting thereon an entire unit with all necessary attachments preassembled at the factory.

Regardless of the type of device employed, certain problems arise, calling for special care. Some of these are described below, with suggestions for their correction:

1. Welding below ground, in cramped quarters and under all-weather conditions, is not conducive to good workmanship. If welding *must* be done, the welder should be an expert pipe welder, and he should have ample room for freedom of movement and ease of visual inspection. A clamp-on, gasketed pitless adapter is easier to install, but requires a smooth and clean surface for the gasket.

2. The pitless unit is manufactured and tested under factory conditions. However, its attachment to the casing may present special problems. If the well casing is threaded and coupled (T&C),

---

[1] National Sanitation Foundation, Post Office Box 1468, Ann Arbor, Mich 48106.
[2] Water Systems Council, 221 North LaSalle St , Chicago, Ill 60601

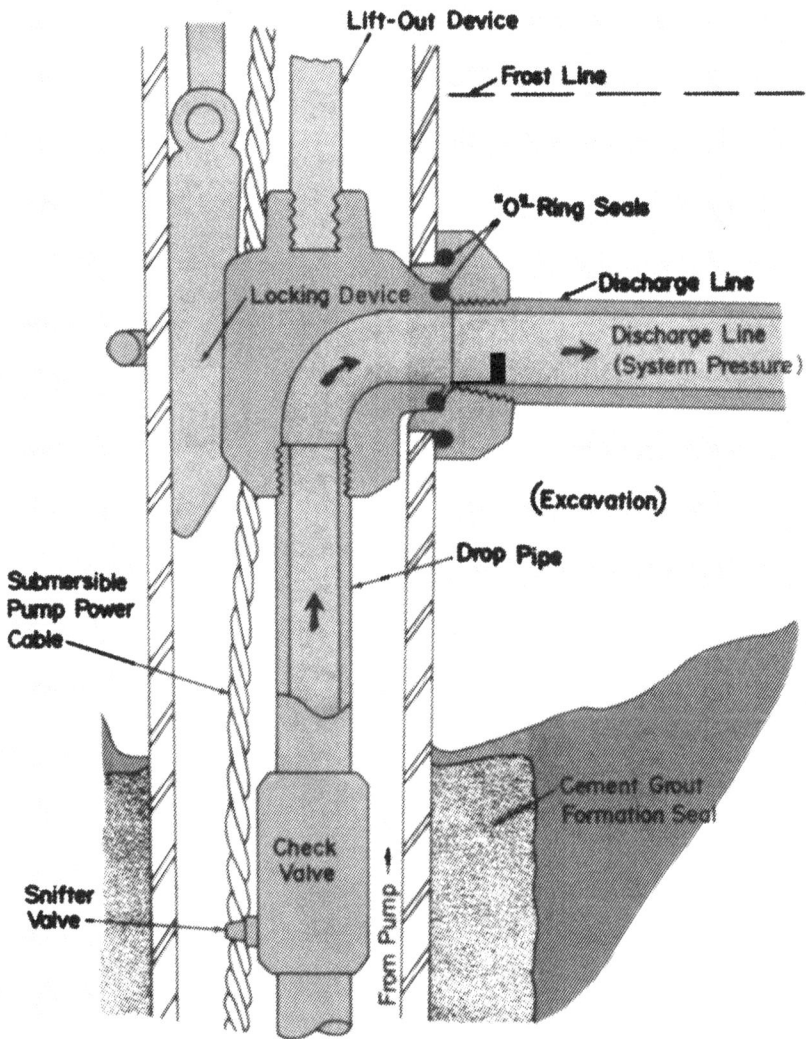

FIGURE 21.   Clamp-on pitless adapter for submersible pump installation.

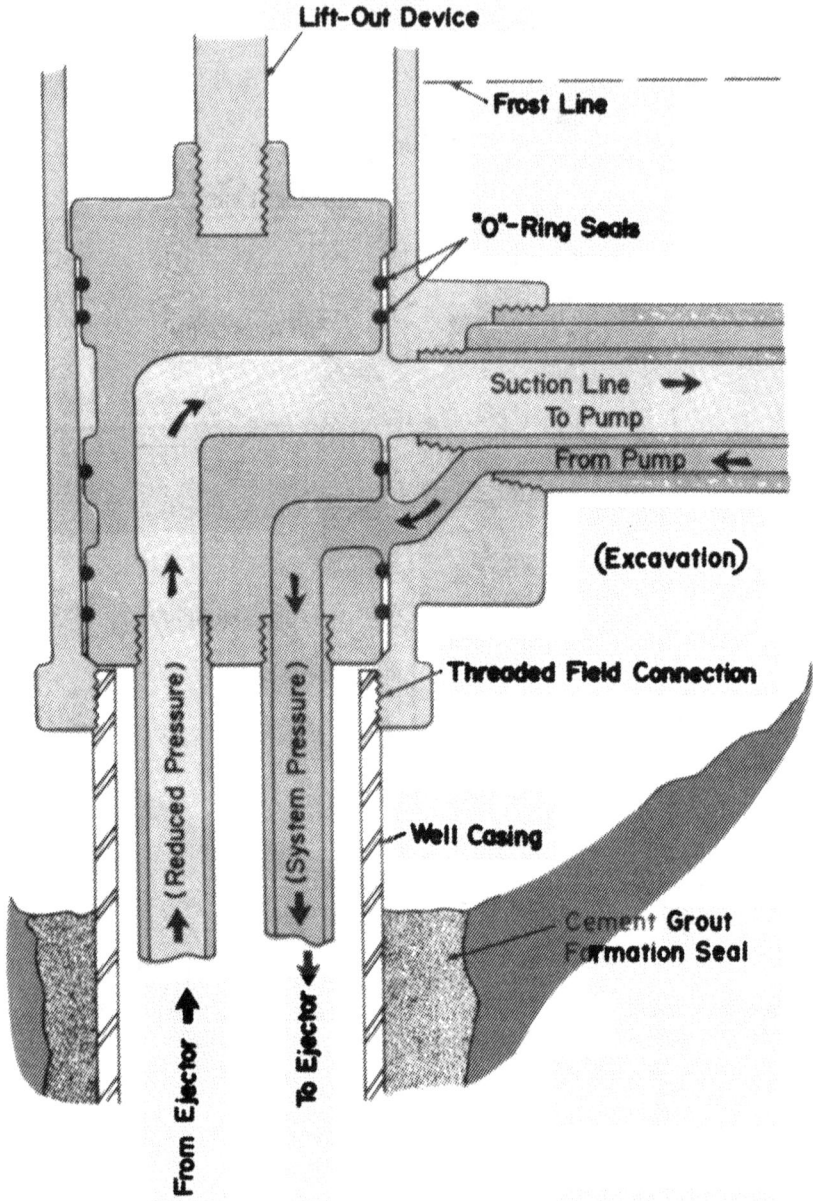

Lift-Out Device

Frost Line

"O"-Ring Seals

Suction Line →
To Pump

From Pump ←

(Excavation)

(Reduced Pressure)

(System Pressure)

Threaded Field Connection

Well Casing

Cement Grout
Formation Seal

From Ejector ↑

To Ejector ↓

FIGURE 22.   Pitless unit with concentric external piping
for jet pump installation.

112

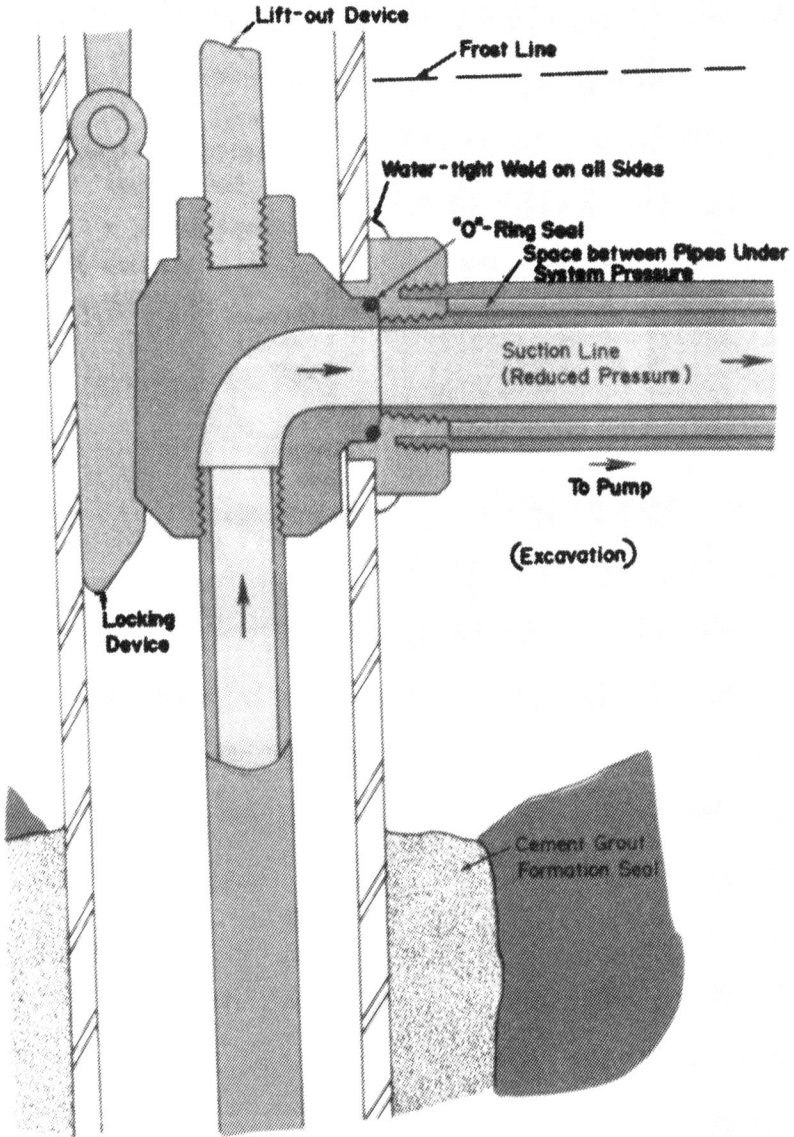

FIGURE 23. Weld-on pitless adapter with concentric external piping for "shallow well" pump installation.

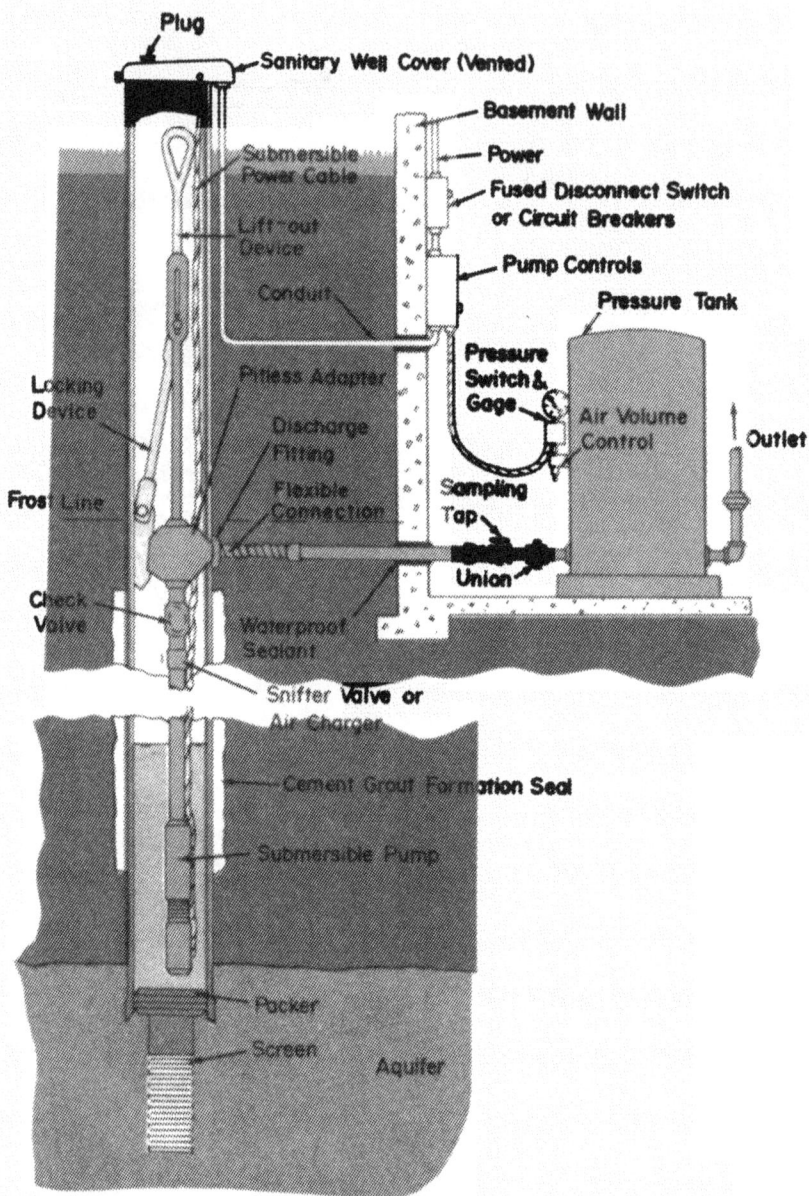

FIGURE 24. Pitless adapter with submersible pump installation for basement storage.

it may be possible to adjust the height of one of the joints so that it is about right for the attachment of the unit. If the height cannot be adjusted, or if welded joints have been made, the casing must be cut off at the proper depth below ground and then threaded.

Power-driven pipe-threading machines can be used to thread casing "in place" in sizes up to and including 4 inches. Between 10 and 12 full threads should be cut on the casing to make a good, strong joint. The threads should be good quality, cut with dies in good condition.

When it is necessary to weld, the first requirement is that the casing be cut off square. This cut can be made by special casing-cutting tools working inside the casing, or by "burning" with an acetylene torch from outside. If the torch method is used, it is better to use a jig that attaches to the casing, supporting and guiding the torch as the casing is burned off.

A competent welder should be able to make a strong weld if given enough room in which to work. It is not so easy to get a watertight joint under these conditions. Two or three "passes" around the pipe should be made, following recommended procedures for pressure pipe welding. The final welded connection should be at least as thick, as strong, and as resistant to corrosion as the well casing itself.

3. Clamps and gaskets are used for attachment of both adapters and units. These devices have been criticized by some health departments because of their relative structural weakness as compared with other connections. It is feared by some that the joint is more easily broken or caused to leak by mechanical damage, or by frost-heave acting on the casing or the well slab.[3]

It is apparent that a watertight joint requires good contact between the gasket and the surfaces against which it is to seal. Corrosion-resistant, machined surfaces provide better conditions for this seal. When the rubber gasket is required to seal against the casing, special care must be taken to assure that the contact surface is clean and smooth. Clamp-and-gasket connections should be designed so that forces resulting from weight, misalinement, twisting, settlement, and vibration are resisted by the metal parts, and not by the rubber gaskets.

4. Materials used in adapters, adapter units, and accessories should be selected carefully for strength and resistance of corrosion. Corrosion potential is high in the earth formations found closest to the surface and where there is moisture and air. To use metals of differing "potential" in contact with each other in a corrosive environment is to invite rapid destruction of one of them by electrolytic corrosion. For example, steel clamps would be more

---

[3] Some States prohibit the use of "Dresser type" connections for pitless units

compatible with steel casing than most other metals or alloys. Some metals that by themselves resist corrosion — e.g., bronze, brass, copper, aluminum — may corrode, or cause others to corrode, when placed in contact with dissimilar metals. Different metals placed in a corrosive environment should be insulated from each other by rubber, plastic, or other nonconductor. Care should be taken in the selection of welding materials; the welded connection is frequently the point where corrosion begins.

Cast iron is more resistant to corrosion than steel under many conditions of soil and water corrosiveness. However, some grades of cast iron are unable to resist severe stresses resulting from tension, bending, and impact. Metals used in castings subjected to such loads should be selected, and the parts designed, to meet these requirements. The consequences of breakage can be serious and expensive, especially if pumping equipment, pipes, and accessories fall into the well.

For the same reasons, plastics should be used in adapters and units only where they are not subjected to severe forces of bending, tension, or shear.

5. Extensive excavation around the well produces unstable soil conditions, and later settlement is to be expected. Settlement of the discharge line, unless at least a portion of the line is flexible, will place a load on the adapter connection that could cause it to break or leak. If for some reason the use of rigid pipe is necessary, the connection should be by means of a "gooseneck," a "swing joint," or other device that will adjust to the settlement without transferring the load to the adapter. The best fill material to use to minimize settlement of the discharge line is fine to medium sand, washed into place. With a correctly placed cement grout seal around the casing and below the point of attachment (see fig. 25), the sand will not find its way into the well. Sand does not shrink or crack in drying, and several feet of it form an efficient barrier against penetration by bacteria.

6. Once a pitless unit has been installed and tested, there remains the risk of accidental damage to the buried connection. Numerous cases of breakage by bulldozers and other vehicles have been documented. Until all construction and grading around the area have been completed, the well should be marked clearly with a post and flag. A "2 x 4" 3 or 4 feet long, clamped or wired securely to the well casing and bearing a red flag, has proved effective.

If the well is located in an area where motor vehicles are likely to be operated, the final installation should include protective pipe posts set in concrete. The posts should be just high enough to protect the well, but not so high that they would interfere with well servicing.

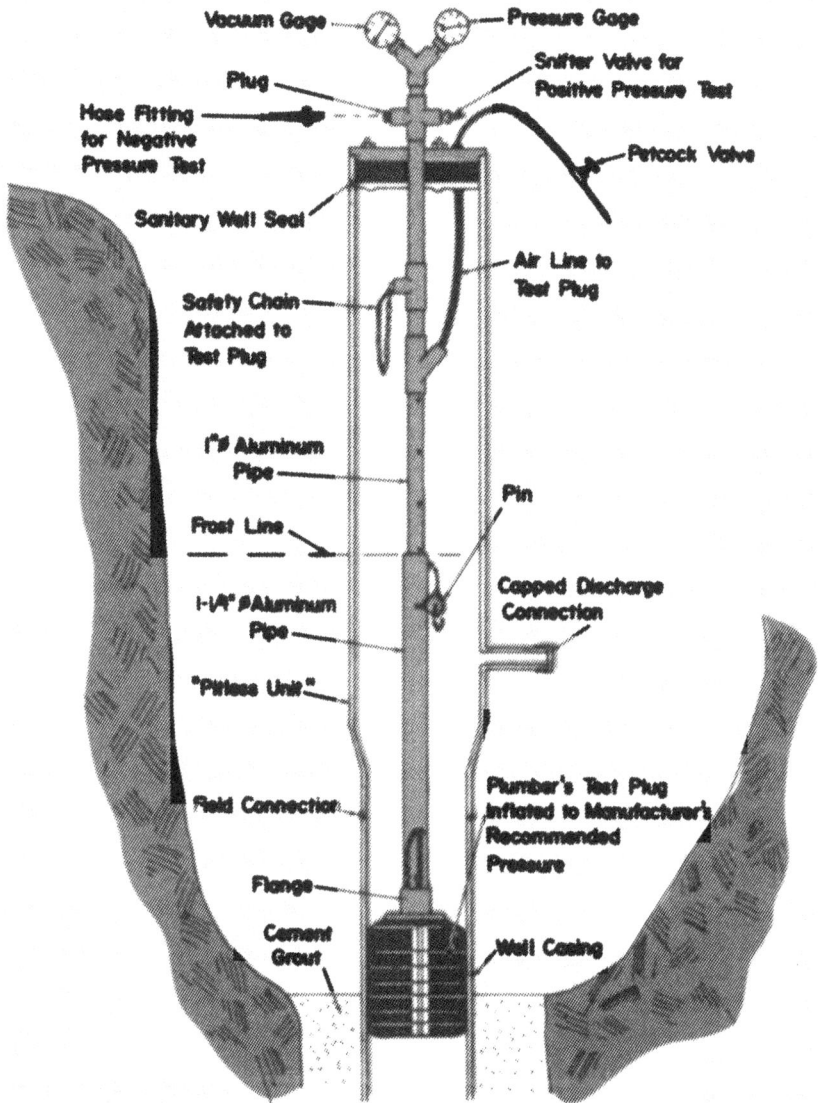

FIGURE 25. Pitless adapter and unit testing equipment.

117

Inspection and Testing of Pitless Devices

Pitless adapters and units are installed within the upper 10 feet of the well structure – the zone of greatest potential for corrosion and contamination. Procedures for inspecting and testing are therefore important.

The buyer should select an adapter or unit that not only satisfies health department requirements and the design criteria above, but whose manufacturer will stand behind it.

Hiring a contractor with a reputation for good work is probably the best assurance of getting the job done right. The owner should insist that the contractor guarantee his work for at least 1 year. Some State and local health departments maintain lists of licensed or certified contractors authorized by law to construct wells and install pumping systems.

Field connections on pitless adapters and units can be easily tested with the equipment shown in figure 25. The lower plug is first positioned just below the deepest joint to be tested, and then inflated to the required pressure. The sanitary well seal is then positioned in the top of the well and tightened securely to form an airtight seal. This isolated section of the casing or unit is then pressurized through the discharge fitting, or through a fitting in the sanitary well seal. (See fig. 25.) A pressure of 5 to 7 pounds per square inch should be applied and this pressure maintained, without the addition of more air, for 1 hour. *Warning* Do not hold face over well seal while pressurized! While under pressure, all field connections should be tested for leaks with soap foam. Any sign of leakage – either by loss of pressure or by the appearance of bubbles through the soap – calls for repair and retesting.

Adapters and units that depend on rubber or plastic seals in the field connection should also be tested under *negative* pressure conditions. This can be accomplished by connecting the hose fitting (fig. 25) to a source of vacuum. The negative pressure is read on the vacuum gage.

Positive pressure may be applied to the isolated section by means of a tire pump, but a powered source makes the job much easier and encourages better testing. If an air compressor is not available or handy, a tire-inflation kit of the kind that uses automobile engine compression will be found convenient. *The plumber's test plug should only be inflated by means of a hand-operated tire pump.*

Negative pressure is most readily applied by connecting a length of vacuum hose (heavy wall, small bore) between the hose fitting in the well seal and the vacuum system of an automobile engine. To reach the desired negative pressure range (10 to 14 inches of mercury vacuum), it may be necessary to accelerate the engine for a period of time. Once the desired range is reached, the hose is clamped shut or plugged, the engine disconnected, and the vacuum gage observed over a period of 1 hour to see whether there is any detectable loss of negative pressure.

Leaks found in rubber or plastic seals should be closed by tightening the clamps, if possible. In case a cement sealant must be applied, it should be of a kind that will provide a strong yet flexible bond between the sealing surfaces, and should be compounded to provide long service when buried.

A negative pressure of 10 inches of mercury vacuum is equivalent to about 11.3 feet of flood water over the joint in question when the well casing is at atmospheric pressure.

DISTRIBUTION

Pipe and Fittings

For reasons of economy and ease of construction, distribution lines for small water systems are ordinarily made up with standard threaded, galvanized iron or steel pipe and fittings. Other types of pipes used are cast iron, asbestos-cement, concrete, plastic, and copper. Under certain conditions and in certain areas, it may be necessary to use protective coatings, galvanizing, or have the pipes dipped or wrapped. When corrosive water or soil is encountered, copper, brass, wrought iron, plastic or cast iron pipe, although usually more expensive initially, will have a longer, more useful life. Cast iron is not usually available in sizes below 2 inches in diameter; hence, its use is restricted to the larger transmission lines.

Plastic pipe for cold water piping is usually simple to install, has a low initial cost, and has good hydraulic properties. When used in a domestic water system, plastic pipe should be certified by an acceptable testing laboratory (such as the National Sanitation Foundation) as being nontoxic and non-taste-producing. It should be protected against crushing and from attack by rodents. Asbestos-cement pipe for water systems, available in the sizes required, has the advantages of ease of installation and moderate resistance to corrosion.

Fittings are usually available in the same sizes and materials as piping, but valves are generally cast in bronze or other alloys. In certain soils the use of dissimilar metals in fittings and pipe may create electrolytic corrosion problems. The use of nonconductive plastic inserts between pipe and fittings or the installation of sacrificial anodes is helpful in minimizing such corrosion.

Pipes should be laid as straight as possible in trenches, with air-relief valves or hydrants located at the high points on the line. Failure to provide for the release of accumulated air in a pipeline on hilly ground may greatly reduce the capacity of the line. It is necessary that pipeline trenches be excavated deep enough to prevent freezing in the winter. Pipes placed in trenches at a depth of more than 3 feet will help to keep the water in the pipeline cool during the summer months.

Pipe Capacity and Head Loss

The pipeline selected should be adequate to deliver the required peak flow of water without excessive loss of head; i.e., without decreasing the discharge pressure below a desirable minimum. The normal operating water pressure for household or domestic use ranges from 20 to 60 pounds per square inch,[4] or about 45 to 140 feet of head at the fixture.

The capacity of a pipeline is determined by its size, length, and interior surface condition. Assuming that the length of the pipe is fixed and its interior condition established by the type of material, the usual problem in design of a pipeline is that of determining the required diameter.

The correct pipe size can be selected with the aid of figure 26, which gives size as a function of head loss, H, length of pipeline, L, and peak discharge, Q. As an example of the use of figure 26, suppose that a home and farm installation is served by a reservoir a minimum distance of 500 feet from the point of use, one whose surface elevation is at least 150 feet above the level of domestic service, and in which a minimum service pressure of 30 pounds per square inch is required. It will be necessary first to determine the maximum operating head loss, i.e., the difference in total head and the required pressure head at the service.

$$H = 150 - 2.3 \times 30 = 150 - 69 = 81 \text{ feet}$$

The maximum peak demand which must be delivered by the pipeline is determined to be 30 gallons per minute.

$$Q = 30 \text{ gallons per minute}$$

The hydraulic gradient is 0.162 foot per foot.

$$\frac{H}{L} = \frac{81}{500} = 0.162 \text{ foot per foot}$$

Entering figure 26, with the computed values of H/L and Q, one finds that the required standard galvanized pipe size is approximately 1-3/8 inches. Since pipes are available only in standard dimensions, standard pipe of 1½ inches in diameter (the next size) should be used.

Additional head losses may be expected from the inclusion of

---

[4]One pound per square inch is the pressure produced by a column of water 2 31 feet high.

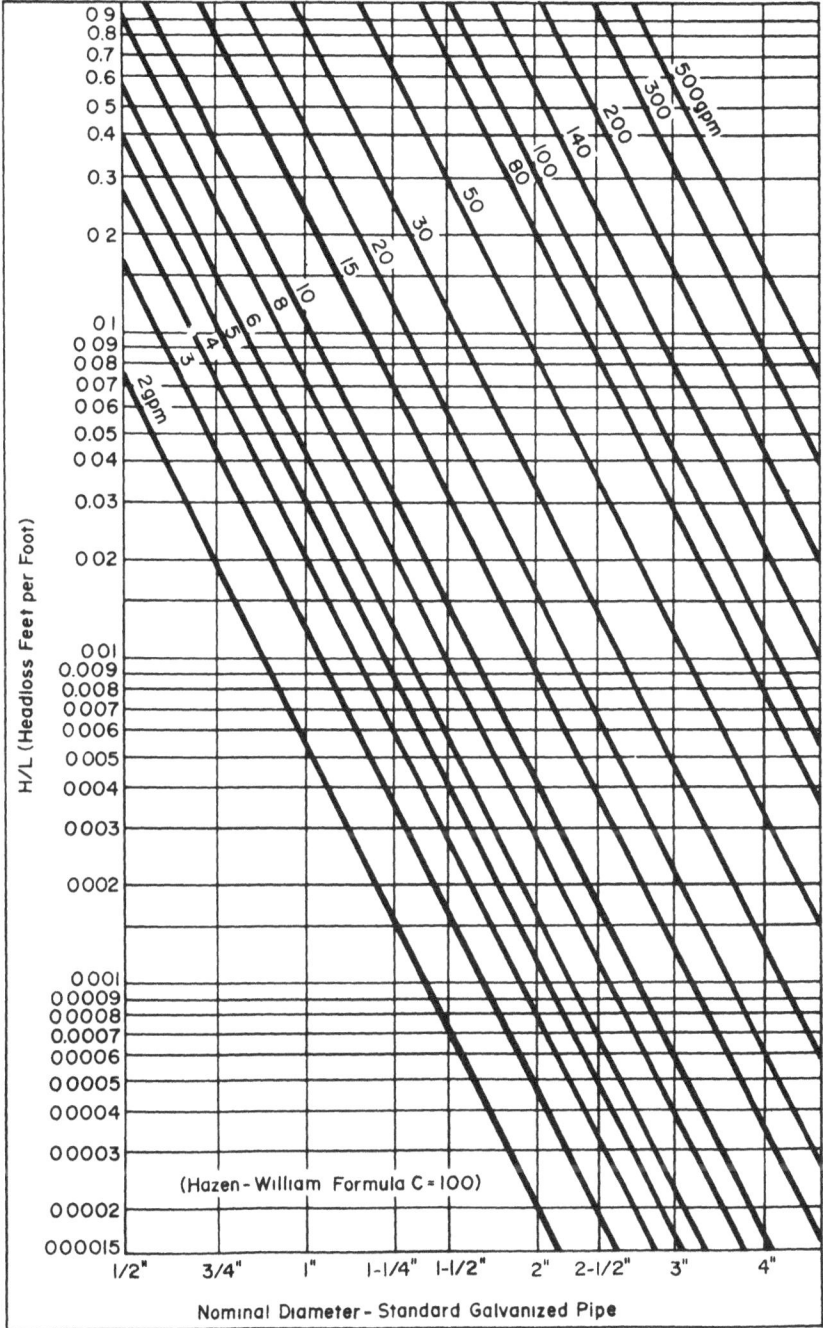

FIGURE 26. Head loss versus pipe size.

121

fittings in the pipeline. These losses may be expressed in terms of the equivalent to the length and size of pipe which would produce an equivalent loss if, instead of adding fittings, we added additional pipe. Table 8 lists some common fitting losses in terms of an equivalent pipe length.

TABLE 8. — *Allowance in equivalent length of pipe for friction loss in valves and threaded fittings*

Diameter of fitting	90° std. ell	45° std. ell	90° side tee	Coupling or straight run	Gate valve	Globe valve	Angle valve
Inches	*Feet*	*Feet*	*Feet*	*Feet*	*Feet*	*Feet*	*Feet*
3/8	1	0.6	1.5	0.3	0.2	8	4
1/2	2	1.2	3	0.6	0.4	15	8
3/4	2.5	1.5	4	0.8	0.5	20	12
1	3	1.8	5	0.9	0.6	25	15
1-1/4	4	2.4	6	1.2	0.8	35	18
1-1/2	5	3	7	1.5	1.0	45	22
2	7	4	10	2	1 3	55	28
2-1/2	8	5	12	2.5	1.6	65	34
3	10	6	15	3	2	80	40
3-1/2	12	7	18	3.6	2.4	100	50
4	14	8	21	4	2.7	125	55
5	17	10	25	5	3.3	140	70
6	20	12	30	6	4	165	80

In the example given above the inclusion of two gate valves (open), two standard elbows, and two standard tees (through) would produce a head loss equivalent to 15 feet of 1½-inch pipe. From figure 26 one finds that by using 515 feet of 1½-inch pipe instead of the actual length of 500 feet (H/L=0.157), the capacity of the system for the same total head loss is about 38 gallons per minute, a satisfactory discharge.

It can be seen from this example that fitting losses are not particularly important for fairly long pipelines, say greater than about 300 feet. For pipelines less than 300 feet, fitting losses are very important and have a direct bearing on pipe selected; therefore, they should be calculated carefully.

Globe valves which do produce large head losses should be avoided in main transmission lines for small water systems.

Interior piping, fittings, and accessories should conform to the minimum requirements for plumbing of the *National Plumbing Code*[5] or equivalent applicable plumbing code of the locality.

Protection of Distribution Systems

The sanitary protection of new or repaired pipelines can be facilitated by proper attention to certain details of construction. All

---

[5]Obtainable at the American Society of Mechanical Engineers, United Engineering Center, 345 East 47th St., New York, N.Y. 10017.

connections should be made under dry conditions, either in a dry trench or, if it is not possible to completely dewater the trench, above the ground surface. Soiled piping should be thoroughly cleaned and disinfected before connections are made. Flush valves or cleanouts should be provided at low points where there is no possibility of flooding.

When not properly designed or installed, frostproof hydrants may permit contamination to enter the water system. Such hydrants should be provided with suitable drainage to a free atmosphere outlet where possible. The drainage from the base of the hydrant should not be connected to a seepage pit which is subject to pollution or to a sewer. The water-supply inlet to water tanks used for stock, laundry tubs, and other similar installations should be placed with an air gap (twice pipe diameter) above the flooding level of the fixtures to prevent danger of back siphonage. There should be no cross-connection, auxiliary intake, bypass, or other piping arrangement whereby polluted water or water of questionable quality can be discharged or drawn into the domestic water supply system.

Before a distribution system is placed in service it should be completely flushed and disinfected.

Disinfection of Water-Distribution System

General

These instructions cover the disinfection of water distribution systems and attendant standpipes or tanks. It is always necessary to disinfect a water system before placing it in use under the following conditions:

1. Disinfection of a system that has been in service with raw or polluted water, preparatory to transferring the service to treated water.
2. Disinfection of a new system upon completion and preparatory to placing in operation with treated water or water of satisfactory quality.
3. Disinfection of a system after completion of maintenance and repair operations.

Procedure

The entire system, including tank or standpipe, should be thoroughly flushed with water to remove any sediment that may have collected during operation with raw water. Following flushing, the system should be filled with a disinfecting solution of calcium hypochlorite and treated water. This solution is prepared by adding 1.2 pounds of high-test 70 percent calcium hypochlorite to each 1,000 gallons of water, or by adding 2 gallons of ordinary household liquid bleach to each 1,000 gallons of water. A mixture of this kind provides a solution having not less than 100 mg/ℓ of available chlorine.

The disinfectant should be retained in the system, tank, or standpipe, if included, for not less than 24 hours, then examined for residual chlorine and drained out. If no residual chlorine is found present, the process should be repeated. The system is next flushed with treated water and put into operation.

STORAGE

Determination of Storage Volume

Three types of storage facilities are commonly employed for individual water supply systems. They are pressure tanks, elevated storage tanks, and ground-level reservoirs and cisterns.

When ground water sources with sufficient capacity and not requiring treatment are used, only a small artificial storage facility may be needed since the water-bearing formation tapped constitutes a natural storage area.

Pressure Tanks. Pressure in a distribution system served by a pneumatic tank is maintained by pumping water directly to the tank from the source. This pumping action compresses a volume of entrapped air. The air pressure equal to the water pressure in the tank can be controlled between desired limits by means of pressure switches which stop the pump at the maximum setting and start it at the minimum setting. *The capacity of pressure tanks is usually small when compared to the total daily water consumption.* Tanks are designed to meet only peak demands because only 10 to 40 percent of tank volume is usable storage. The maximum steady demand the system can deliver is equal to the pump capacity.

The usable storage of a pressure tank can be increased by "supercharging" with air at the time of installation, or by precharging at the factory. Precharging can only be done in tanks having the water space and air space completely separated by a diaphragm or bladder. Consult your dealer for design details and characteristics.

The Water Systems Council [6] recommends the figures in Tables 9 and 10 for the selection of pumps and pressure tanks for various size homes.

TABLE 9 *Seven-minute Peak Demand Period Usage*

Number of baths in home:	1	1½	2–2½	3–4
Normal 7-min peak demand (gal.)	45	75	98	122
Minimum size pump to meet peak demand without using storage	7 GPM	10 GPM	14 GPM	17 GPM

Note· Values given are average and do not include higher or lower extremes.

---

[6] Water Systems Council, 221 North LaSalle Street, Chicago, IL 60601

Using the pump capacity obtained from Table 9, find the tank size that corresponds to the kind of tank (precharged, supercharged, or plain) and the pressure range:

TABLE 10. *Tank Selection Chart—Gallons (Based on present industry practice)*

| PUMP CAPACITY | | Minimum Draw- | Switch Setting Pounds Per Square Inch) | | | | | | | | |
| | | | 20-40 | | | 30-50 | | | 40-60 | | |
GPH	GPM	Down (Gals)	A*	B*	C*	A*	B*	C*	A*	B*	C*
240	4	4	15	15	20	15	20	30	20	20	40
300	5	5	15	20	30	20	25	40	25	25	50
360	6	6	20	20	35	25	25	45	30	30	55
420	7	7	20	25	40	25	30	55	30	40	75
480	8	8	25	30	40	30	35	65	35	45	85
540	9	9	30	30	50	35	40	70	40	50	95
600	10	10	30	35	55	40	45	80	45	55	105
660	11	12	35	40	60	45	50	95	55	65	125
720	12	13	40	45	70	50	60	105	60	70	135
780	13	15	45	50	80	60	65	120	70	80	155
840	14	17	55	60	90	65	75	135	75	90	175
900	15	19	60	65	100	75	80	150	85	105	195
960	16	20	65	70	110	75	90	160	95	115	205
1020	17	23	70	80	120	90	100	185	105	125	240
1080	18	25	80	85	135	95	110	200	115	140	260
1140	19	27	85	95	150	105	120	215	125	150	280
1200	20	30	95	105	160	115	130	240	140	165	310

*A—Precharged bladder or diaphragm tank   B—Supercharged floating water tank   C—Plain steel tank

the minimum tank size needed.

When a pressure tank is provided in the distribution system there is no difficulty with water hammer. Otherwise, it may be necessary to provide an air chamber on the discharge line from the well located near the pump to minimize the effects caused by water hammer.

Elevated Storage. Elevated tanks should have a capacity which is at least equal to 2 days' average consumption requirement. Larger storage volume may be necessary to meet special demands such as firefighting or equipment cleanup operations.

Ground-Level Reservoirs and Cisterns Reservoirs that receive surface runoff should generally be large enough to supply the average daily demand over a drought period of maximum length. Cisterns are customarily designed with sufficient capacity to provide water during periods less than 1 year in duration.

Protection of Storage Facilities

Suitable storage facilities for relatively small systems may be constructed of concrete, steel, brick, and sometimes of wood above the land surface, or of concrete or brick if partially or wholly below the ground surface. Such storage installations should receive the same care as cistern installations in the selection of a suitable location and provision against contamination. Asphalt or tar for waterproofing the interior of storage units is not recommended because of the objectionable taste imparted to the water and the possibility of undesirable chemical reaction with the materials used for treatment. Specifications covering the painting of water tanks are available from the American Water Works Association.[7] Appropriate Federal, State, or local health agencies should be consulted relative to approved paint coatings for interior tank use.

All storage tanks for domestic water supply should be completely covered and so constructed as to prevent the possibility of pollution of the tank contents by outside water or other foreign matter. Figures 27 and 28 show some details for manhole covers and piping connections to prevent the entrance of pollution from surface drainage. Concrete and brick tanks should be made watertight by a lining of rich cement mortar. Wood tanks are generally constructed of redwood or cypress and while filled they will remain watertight. All tanks require adequate screening of any openings to protect against the entrance of small animals, mosquitoes, flies, and other small insects.

Tanks containing water to be used for livestock should be partially covered and so constructed that cattle will not enter the tank. The area around the tank should be sloped to drain away from the tank.

Figure 27 shows a typical concrete reservoir with screened inlet and outlet pipes. This figure also illustrates the sanitary manhole cover. The cover should overlap by at least 2 inches a rim elevated at least 4 inches to prevent drainage from entering the reservoir. This type of manhole frame and cover should be designed so that it may be locked to prevent access by unauthorized persons.

The water in storage tanks, cisterns, or pipelines should not be polluted with an emergency water supply that has been polluted at its source or in transit.

Disinfection of storage facilities subsequent to construction or repair should be carried out in accordance with the recommendations stated under "Disinfection of Water Distribution System" in this part of the manual.

---

[7] American Water Works Association, 2 Park Ave , New York, N.Y. 10016

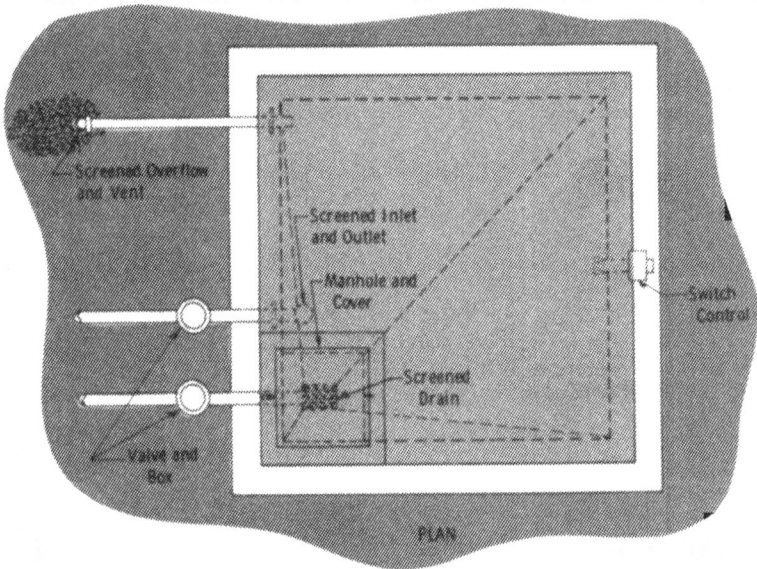

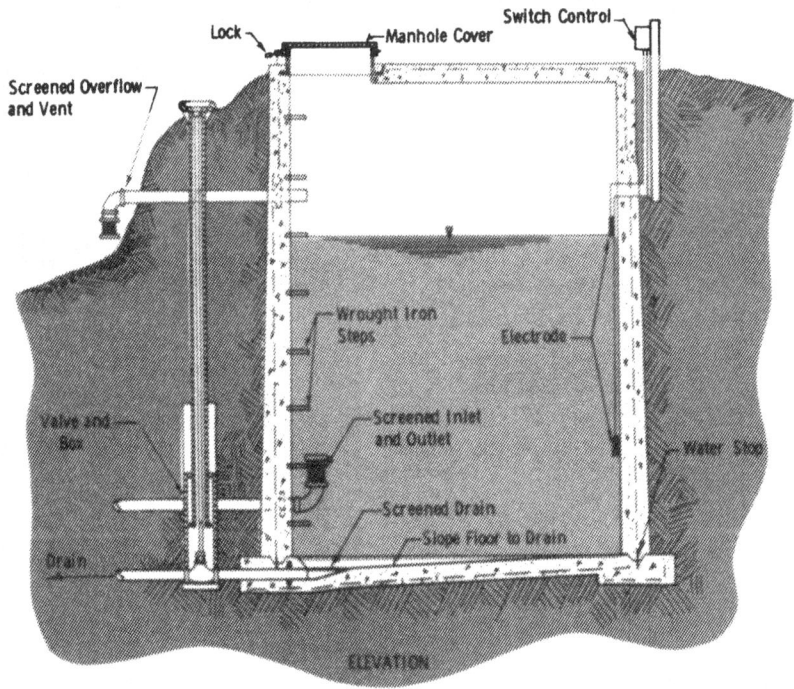

FIGURE 27. Typical concrete reservoir.

Overlapping, Circular Iron Cover

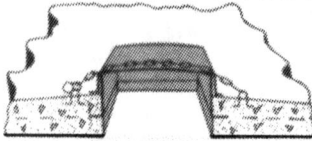

Iron Cover

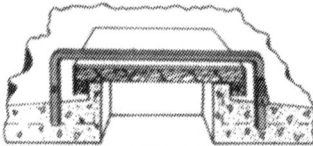

Galvanized Sheet Metal Over Wooden Cover

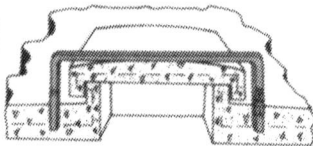

Concrete Cover

MANHOLE COVERS

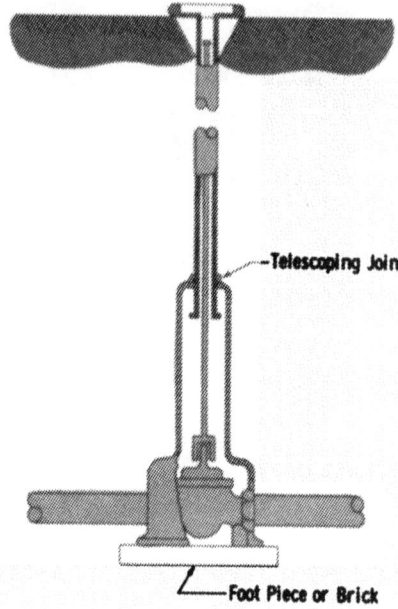

Telescoping Joint

Foot Piece or Brick

TYPICAL VALVE AND BOX

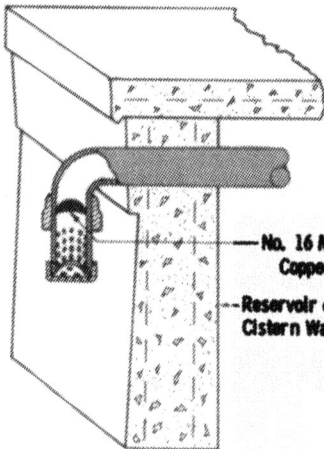

Coupling

Pipe Connection With Anchor Flange Casting

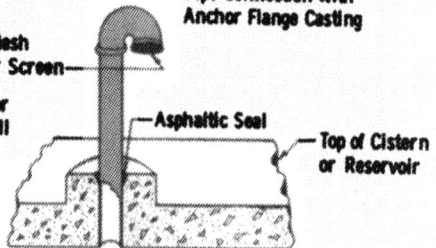

No. 16 Mesh Copper Screen

Reservoir or Cistern Wall

Asphaltic Seal

Top of Cistern or Reservoir

OVERFLOW AND VENT

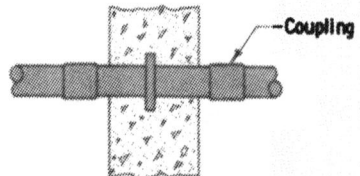

VENT

FIGURE 28. Typical valve and box, manhole covers, and piping installations.

Bibliography

List of References on Individual Water Supply Systems

American Public Health Association, American Water Works Association, and Water Pollution Control Federation, *Standard Methods for the Examination of Water and Waste Water,* 15th ed., Amer. Pub. Hlth. Assn., Washington, D.C. (1975).

American Water Works Association, *Water Quality and Treatment,* 3d ed , American Water Works Association, New York, N Y. (1971)

American Water Works Association, American Society of Civil Engineers, and Conference of State Sanitary Engineers, *Water Treatment Plant Design,* American Water Works Association, New York, N.Y. (1969)

American Water Works Association, Committee on Viruses in Water, "Viruses in Water," *Journal of the American Water Works Association,* Vol 61, No 10, pp. 491-494 (1969).

Anderson, Keith E., *Water Well Handbook,* Missouri Water Well and Pump Contractors Association. Rolla. Mo (1971).

Baker, R. J., Carroll, L. J., and Laubusch, E J , *Water Chlorination Handbook,* American Water Works Association, New York, N.Y. (1972).

Capitol Controls Co., "Chlorination Guide," Capitol Controls Co., Colmar, Pa (undated)

Chang, S. L , "Iodination of Water," *Boletín de la Oficina Sanitaria Panamericana,* Vol. 59, pp. 317-331 (1966)

Departments of the Army and the Air Force, *Well Drilling Operations* (TM 5-297, AFM 85-23), U.S Government Printing Office, Washington, D.C. (Sept 1965)

Gibson, U. P , and Singer, R. D., *Water Well Manual,* Premier Press, Berkeley, Calif. (1971)

Goldstein, Melvin, McCabe, L. J., Jr., and Woodward, Richard L., "Continuous-Flow Water Pasteurizer for Small Supplies," *Journal of the American Water Works Association,* Vol 52, No. 2. pp. 247-254 (Feb. 1960)

Hill, R. D., and Schwab, G. O., "Pressurized Filters for Pond Water Treatment." *Transactions of the ASAE,* Vol. 7, No. 4, pp. 370-374, 379, American Society of Agricultural Engineers. St Joseph, Mich. (1964).

Hodgkinson, Carl, *Removal of Coliform Bacteria from Sewage by Percolation through Soil*, University of California, Sanitary Engineering Research Laboratory, IER Series 90, No. 1, Berkeley, Calif. (1955).

Hooker, Dan, "How to Protect the Submersible Pump from Lightning Surge Damage," Bulletin DPED-27, General Electric Co., Pittsfield, Mass. (1969)

Inter-Agency Committee on Water Resources, "Inventory of Federal Sources of Ground Water Data," *Notes on Hydrologic Activities*, Bulletin No 12, U.S. Geological Survey, Washington, D.C. (1966)

Joint Committee on Plastics. *Thermoplastic Materials, Pipe, Fittings, Valves, Traps, and Joining Materials (Standard No. 14)*, National Sanitation Foundation, Ann Arbor, Mich. (1965).

National Association of Plumbing, Heating, and Cooling Contractors, *National Standard Plumbing Code*, National Association of Plumbing, Heating, and Cooling Contractors, Washington, D.C. (1971).

National Fire Protection Association, "Water Supply Systems for Rural Fire Protection," *National Fire Codes*, Vol. 8, Boston, Mass. (1969).

National Sanitation Foundation Testing Laboratory, *Seal of Approval Listing of Plastic Materials, Pipe, Fittings and Appurtenances for Potable Water and Waste Water*, National Sanitation Foundation, Ann Arbor, Mich. (revised annually).

National Water Well Association, "The Authoritative Primer Ground Water Pollution," *Water Well Journal*, Special Issue, Vol. 24, No. 7 (1970).

Olin Corporation, "Hypochlorination of Water," Olin Corporation – Chemicals Division, New York, N.Y. (1962)

Tardiff, R. D., and McCabe, L. J., "Rural Water Quality Problems and the Need for Improvement," *Second Water Quality Seminar Proceedings,* pp. 34-36, American Society of Agricultural Engineers, St. Joseph, Mich. (1968).

Todd, D. K., *The Water Encyclopedia*, Water Information Center, Manhasset Isle, Port Washington, N.Y (1970).

U. O. P. Johnson Division, *Ground Water and Wells*, U. O. P. Johnson Division, St. Paul, Minn. (1972).

U.S. Department of Health, Education, and Welfare, "A Guide to Reading on Fluoridation," U.S. Public Health Service Pub. No. 1680, Environmental Protection Agency National Environmental Research Center, Cincinnati, Ohio (1970)

U.S. Department of Health, Education, and Welfare, "Environmental Health Guide for Mobilehome Communities, with a Recommended Ordinance," Mobile Homes Manufacturers Association, 6650 N. Northwest Highway, Chicago, Ill. 60631 (1971).

U.S. Department of Health, Education, and Welfare, "Policy Statement on Use of the Ultraviolet Process for Disinfection of Water," Environmental Protection Agency, Water Supply Division, Washington, D.C. (Apr. 1, 1966).

U.S. Department of Housing and Urban Development, "Minimum Property Standards for One and Two Living Units," Federal Housing Administration, U.S. Government Printing Office, Washington, D.C. (1966).

U.S. Department of the Interior, Geological Survey, "A Primer on Ground Water," U.S. Government Printing Office, Washington, D.C. (reprinted annually).

U.S. Department of the Interior, Geological Survey, "A Primer on Water," U.S. Government Printing Office, Washington, D.C. (reprinted annually).

U.S. Department of the Interior, Geological Survey, "A Primer on Water Quality," U.S. Government Printing Office, Washington, D.C. (reprinted annually).

U.S. Environmental Protection Agency and American Water Works Association, "Control of Biological Problems in Water Supplies," Environmental Protection Agency, Water Supply Division, Denver, Colo. (1971).

U.S. Environmental Protection Agency, "Fluoridation Engineering Manual," Environmental Protection Agency, Water Supply Division, Washington, D.C. (1972).

U.S. Environmental Protection Agency, "Health Guidelines for Water and Related Land Resources Planning, Development and Management," Environmental Protection Agency, Water Supply Division, Washington, D.C. (1971).

U.S. Environmental Protection Agency, "List of Publications Concerning Water Supply Problems," Environmental Protection Agency, Water Quality Office, Cincinnati, Ohio (1971).

U.S. Environmental Protection Agency, "Manual for Evaluating Public Drinking Water Supplies," U.S. Public Health Service Pub. No. 1820, U.S. Government Printing Office, Washington, D.C. (1971).

U.S. Environmental Protection Agency, "Sanitary Survey of Drinking Water Systems on Federal Water Resource Developments – A Pilot Study," Washington, D.C. (1971).

Water Systems Council, "Water Systems Handbook, " 6th ed., Water Systems Council, Chicago, Ill. (1977).

Whitsell, W. J., and Hutchinson, G. D., "Seven Danger Signals for Individual Water Supply Systems," *Transactions of the American Society of Agricultural Engineers* (in press). American Society of Agricultural Engineers, St. Joseph, Mich. (1973).

Winton, E. F., "The Health Effects of Nitrates in Water," *Proceedings of the Twelfth Sanitary Engineering Conference, Nitrate and Water Supply Source and Control*, Urbana, Ill. (1970).

Woodward, R. L., "The Significance of Pesticides in Drinking Water," *Journal of the American Water Works Association*, Vol. 52, No. 11, pp. 1367-1372 (1960).

Appendix A

Recommended Procedure for Cement Grouting of Wells for Sanitary Protection[1]

The annular open space on the outside of the well casing is one of the principal avenues through which undesirable water and contamination may gain access to a well. The most satisfactory way of eliminating this hazard is to fill the annular space with neat cement grout. To accomplish this satisfactorily, careful attention should be given to see that:

1. The grout mixture is properly prepared.
2. The grout material is placed in one continuous mass.
3. The grout material is placed upward from the bottom of the space to be grouted.

Neat cement grout should be a mixture of cement and water in the proportion of 1 bag of cement (94 pounds) and 5 to 6 gallons of clean water. Whenever possible, the water content should be kept near the lower limit given. Hydrated lime to the extent of 10 percent of the volume of cement may be added to make the grout mix more fluid and thereby facilitate placement by the pumping equipment. Mixing of cement or cement and hydrated lime with the water must be thorough. Up to 5 percent by weight of bentonite clay may be added to reduce shrinkage.

GROUTING PROCEDURE

The grout mixture must be placed in one continuous mass; hence, before starting the operation, sufficient materials should be on hand and other facilities available to accomplish its placement without interruption.

Restricted passages will result in clogging and failure to complete the grouting operation. The minimum clearance at any point, including couplings, should not be less than 1½ inches. When grouting through the annular space, the grout pipe should not be

---

[1] This information has been taken principally from a pamphlet of the Wisconsin State Board of Health entitled "Method of Cement Grouting for Sanitary Protection of Wells." The subject is discussed in greater detail in that publication. (NOTE Publication is out of print.)

less than 1-inch nominal diameter. As the grout moves upward, it picks up much loose material such as results from caving. Accordingly, it is desirable to waste a suitable quantity of the grout which first emerges from the drill hole.

In grouting a well so that the material will move upward, there are two general procedures that may be followed. The grout pipe may be installed within the well casing or in the annular space between the casing and drill hole if there is sufficient clearance to permit this. In the latter case, the grout pipe is installed in the annular space to within a few inches of the bottom. The grout is pumped through this pipe, discharging into the annular space, and moving upward around the casing, finally overflowing at the land surface. In 3 to 7 days the grout will be set, and the well can be completed and pumping started. A waiting period of only 24 to 36 hours is required if quick-setting cement is used.

When the grout pipe is installed within the well casing, the casing should be supported a few inches above the bottom during grouting to permit grout to flow into the annular space. The well casing is fitted at the bottom with an adapter threaded to receive the grout pipe and a check valve to prevent return of grout inside of the casing. After grout appears at the surface, the casing is lowered to the bottom and the grout pipe is unscrewed immediately and raised a few inches. A suitable quantity of water should then be pumped through it, thereby flushing any remaining grout from it and the casing. The grout pipe is then removed from the well and 3 to 7 days are allowed for setting of the grout. The well is then cleared by drilling out the adapter, check valve, plug, and grout remaining within the well.

A modification of this procedure is the use of the well casing itself to convey the grout to the annular space. The casing is suspended in the drill hole and held several feet off the bottom. A spacer is inserted in the casing. The casing is then capped and connection made from it to grout pump. The estimated quantity of grout, including a suitable allowance for filling of crevices and other voids, is then pumped into the casing. The spacer moves before the grout, in turn forcing the water in the well ahead of it. Arriving at the lower casing terminal, the spacer is forced to the bottom of the drill hole, leaving sufficient clearance to permit flow of grout into the annular space and upward through it.

After the desired amount of grout has been pumped into the casing, the cap is removed and a second spacer is inserted in the casing. The cap is then replaced and a measured volume of water sufficient to fill all but a few feet of the casing is pumped into it. Thus all but a small quantity of the grout is forced from the casing into the annular space. From 3 to 7 days are allowed for setting of the grout. The spacers and grout remaining in the casing and drill hole are then drilled out and the well completed.

If the annular space is to be grouted for only part of the total depth of the well, the grouting can be carried out as directed above when the well reaches the desired depth, and the well can then be drilled deeper by lowering the tools inside of the first casing. In this type of construction, where casings of various sizes telescope within each other, a seal should be placed at the level where the telescoping begins, that is, in the annular space between the two casings. The annular space for grouting between two casings should provide a clearance of at least 1½ inches, and the depth of the seal should be not less than 10 feet.

Appendix B

Bacteriological Quality

SAMPLING

In the event that bacteriological samples must be obtained without technical assistance, it is possible to insure satisfactory results by following these steps carefully.

1. Use a sterile sample bottle provided by the laboratory that will examine the sample.
2. Be very careful so that *nothing* except the water to be analyzed will come in contact with the inside of the bottle or the cap. *Do not rinse the bottle.*
3. Inspect the outside of the faucet. If water leaks around the outside of the faucet, a different sampling point should be selected.
4. Allow the water to run for sufficient time to permit clearing of the service line before the sample of water is collected.
5. When filling the bottle, be sure that the bottle is held so that no water which contacts the hands runs into the bottle.
6. Deliver the sample immediately to the laboratory. If samples cannot be processed within 1 hour, the use of iced coolers for storage of samples during transport is recommended. In no case should the time elapsing between collection and examination exceed 30 hours.

EXAMINATIONS

At the present time there are two methods used for determining the bacteriological quality of a water supply. the multiple-tube fermentation technique and the membrane filter technique.

The multiple-tube fermentation technique for determining the presence of coliform bacteria requires 2 to 4 days to obtain results after the sample is received in the laboratory. It also requires the use of trained personnel and centralized laboratory facilities.

In addition, the membrane filter technique is a standard method for making coliform determinations. This technique permits the examination of a greater number of samples than the multiple-tube test, with increased sensitivity in coliform detection. The most important benefit derived from the use of this technique is that definite results are obtained in 18 to 20 hours, a much shorter time than with the multiple-tube procedure. The membrane filter method also permits field testing with self-contained portable kits that are commercially available. The membrane filter technique may be used in disasters and in emergencies such as those arising from floods or hurricanes, where the time which elapses before results of the examination are available is an important consideration in the prompt initiation of protective measures.

Appendix C

Emergency Disinfection

When ground water is not available and surface water must be used, avoid sources containing floating material or water with a dark color or an odor. The water tank from a surface source should be taken from a point upstream from any inhabited area and dipped, if possible, from below the surface.

When the home water supply system is interrupted by natural or other forms of disaster, limited amounts of water may be obtained by draining the hot water tank or melting ice cubes.

In case of a nuclear attack, surface water should not be used for domestic purposes unless it is first found to be free from excessive radioactive fallout. The usual emergency treatment procedures do not remove such substances. Competent radiological monitoring services as may be available in local areas should be relied upon for this information.

There are two general methods by which small quantities of water can be effectively disinfected. One method is by boiling. It is the most positive method by which water can be made bacterially safe to drink. Another method is chemical treatment. If applied with care, certain chemicals will make most waters free of harmful or pathogenic organisms.

When emergency disinfection is necessary, the physical condition of the water must be considered. The degree of disinfection will be reduced in water that is turbid. Turbid or colored water should be filtered through clean cloths or allowed to settle, and the clean water drawn off before disinfection. Water prepared for disinfection should be stored only in clean, tightly covered, noncorrodible containers.

METHODS OF EMERGENCY DISINFECTION

1. *Boiling.* Vigorous boiling for 1 *full* minute will kill any disease-causing bacteria present in water. The flat taste of boiled water can be improved by pouring it back and forth from one container into another, by allowing it to stand for a few hours, or by adding a small pinch of salt for each quart of water boiled.

2. *Chemical Treatment.* When boiling is not practical, chemical disinfection should be used. The two chemicals commonly used are chlorine and iodine.

    a. *Chlorine*

    (1) *Chlorine Bleach.* Common household bleach contains a chlorine compound that will disinfect water. The procedure to be followed is usually written on the label. When the necessary procedure is not given, one should find the percentage of available chlorine on the label and use the information in the following tabulation as a guide:

Available chlorine[1]	Drops per quart of clear water[2]
1%	10
4-6%	2
7-10%	1

[1] If strength is unknown, add 10 drops per quart to purify.
[2] Double amount for turbid or colored water.

The treated water should be mixed thoroughly and allowed to stand for 30 minutes. The water should have a slight chlorine odor; if not, repeat the dosage and allow the water to stand for an additional 15 minutes. If the treated water has too strong a chlorine taste, it can be made more palatable by allowing the water to stand exposed to the air for a few hours or by pouring it from one clean container to another several times.

    (2) *Granular Calcium Hypochlorite.* Add and dissolve one heaping teaspoon of high-test granular calcium hypochlorite (approximately 1/4 ounce) for each 2 gallons of water. This mixture will produce a stock chlorine solution of approximately 500 mg/ℓ, since the calcium hypochlorite has an available chlorine equal to 70 percent of its weight. To disinfect water, add the chlorine solution in the ratio of one part of chlorine solution to each 100 parts of water to be treated. This is roughly equal to adding 1 pint (16 oz.) of stock chlorine solution to each 12.5 gallons of water to be disinfected. To remove any objectionable chlorine odor, aerate the water as described above.

    (3) *Chlorine Tablets.* Chlorine tablets containing the necessary dosage for drinking water disinfection can be purchased in a commercially prepared form. These tablets are available from drug and sporting goods stores and should be used as stated in the instructions. When instructions are not available, use one tablet for each quart of water to be purified.

b. *Iodine*

(1) *Tincture of Iodine.*   Common household iodine from the medicine chest or first aid package may be used to disinfect water. Add five drops of 2 percent United States Pharmacopeia (U.S.P.) tincture of iodine to each quart of clear water. For turbid water add 10 drops and let the solution stand for at least 30 minutes.

(2) *Iodine Tablets.* Commercially prepared iodine tablets containing the necessary dosage for drinking water disinfection can be purchased at drug and sporting goods stores. They should be used as stated in the instructions. When instructions are not available, use one tablet for each quart of water to be purified.

Water to be used for drinking, cooking, making any prepared drink, or brushing the teeth should be properly disinfected.

Appendix D

Suggested Ordinance

The following is suggested for consideration in drafting an ordinance for local application, subject to the approval of the appropriate legal authority, to permit the exercise of appropriate legal controls over nonpublic ground water supply systems used for domestic purposes, to assure that the quality of such water is protected by the proper construction and installation of wells, pumping equipment and appurtenant pipelines.

This suggested legislation has been adapted from U.S. Public Health Service Publication No. 1451, "Recommended State Legislation and Regulations" (July 1965).

Persons using this draft as a guide are urged to familiarize themselves with applicable legal requirements governing the adoption of ordinances of this kind and to adapt the suggested language as may be necessary to meet such requirements.

SUGGESTED LEGISLATION FOR WATER WELL CONSTRUCTION AND PUMP INSTALLATION

[Title should conform to State requirements]

Be it enacted, etc.

Section 1. Short Title

This Act shall be known and may be cited as the "[State] Water Well Construction and Pump Installation Act."

Section 2. Findings and Policy

The [State] legislature finds that improperly constructed, operated, maintained, or abandoned water wells and improperly installed pumps and pumping equipment can affect the public health adversely. Consistent with the duty to safeguard the public health of this State, it is declared to be the policy of this State to require that the location, construction, repair, and abandonment of water wells, and the installation and repair of pumps and pumping equipment conform to such reasonable requirements as may be necessary to protect the public health.

Section 3. Definitions

As used in this act:

(a) *Abandoned water well* means a well whose use has been permanently discontinued. Any well shall be deemed abandoned that is in such a state of disrepair that continued use for the purpose of obtaining ground water is impracticable.

(b) *Construction of water wells* means all acts necessary to obtain ground water by wells, including the location and excavation of the well, but excluding the installation of pumps and pumping equipment.

(c) *Department* means the [designated agency presently having authority to regulate sanitary practices within the State, usually the State department of health].

(d) *Ground water* means water occurring naturally in underground formations that are saturated with water.

(e) *Installation of pumps and pumping equipment* means the procedure employed in the placement and preparation for operation of pumps and pumping equipment, including all construction involved in making entrance to the well and establishing seals, but not including repairs, as defined in this section, to existing installations.

(f) *Municipality* means a city, town, borough, county, parish, district, or other public body created by or pursuant to State law, or any combination thereof acting cooperatively or jointly.

(g) *Pumps* and *pumping equipment* mean any equipment or materials used or intended for use in withdrawing or obtaining ground water, including, without limitation, seals and tanks, together with fittings and controls.

(h) *Pump installation contractor* means any person, firm, or corporation engaged in the business of installing or repairing pumps and pumping equipment.

(i) *Repair* means any action that results in a breaking or opening of the well seal or replacement of a pump.

(j) *Well* means any excavation that is drilled, cored, bored, washed, driven, dug, jetted, or otherwise constructed when the intended use of such excavation is for the location, extraction, or artificial recharge of ground water; but such term does not include an excavation made for the purpose of obtaining or for prospecting tor oil, natural gas, minerals, or products of mining or quarrying, or for inserting media to repressure oil or natural gas bearing formation or for storing petroleum, natural gas, or other products.[1]

(k) *Water well contractor* means any person, firm, or corporation engaged in the business of constructing water wells.

---

[1]Some States may wish to include within the coverage of this definition seismological, geophysical, prospecting, observation, or test wells.

(1) *Well seal* means an approved arrangement or device used to cap a well or to establish and maintain a junction between the casing or curbing of a well and the piping or equipment installed therein, the purpose or function of which is to prevent pollutants from entering the well at the upper terminal.

Section 4. Scope

No person shall construct, repair, or abandon, or cause to be constructed, repaired, or abandoned, any water well, nor shall any person install, repair, or cause to be installed or repaired, any pump or pumping equipment contrary to the provisions of this act and applicable rules and regulations, provided that this act shall not apply to any distribution of water beyond the point of discharge from the storage or pressure tank, or beyond the point of discharge from the pump if no tank is employed, nor to wells used or intended to be used as a source of water supply for municipal water supply systems, nor to any well, pump, or other equipment used temporarily for dewatering purposes.

Section 5. Authority to Adopt Rules, Regulations, and Procedures

The Department shall adopt, and from time to time amend, rules and regulations governing the location, construction, repair, and abandonment of water wells, and the installation, and repair of pumps and pumping equipment, and shall be responsible for the administration of this act. With respect thereto it shall:

(a) Hold public hearings, upon not less than sixty (60) days' prior notice published in one or more newspapers, as may be necessary to assure general circulation throughout the State, in connection with proposed rules and regulations and amendments thereto.[2]

(b) Enforce the provisions of this act and any rules and regulations adopted pursuant thereto.

(c) Delegate, at its discretion, to any municipality any of its authority under this act in the administration of the rules and regulations adopted hereunder.

(d) Establish procedures and forms for the submission, review, approval, and rejection of applications, notifications, and reports required under this act.

(e) Issue such additional regulations, and take such other actions as may be necessary to carry out the provisions of this act.

---

[2]This requirement should be consistent with the general practice for publication requirements in the State and with any State administrative procedure act that may apply.

Section 6. Prior Permission and Notification

(a) Prior permission shall be obtained from the Department for each of the following:

(1) The construction of any water well

(2) The abandonment of any water well

(3) The first installation of any pump or pumping equipment in any well

in any geographical area where the Department determines such permission to be reasonably necessary to protect the public health, taking into consideration other applicable State laws, provided that in any area where undue hardship might arise by reason of such requirement, prior permission will not be required.

(b) The Department shall be *notified* of any of the following whenever prior permission is not required:

(1) The construction of any water well

(2) The abandonment of any water well

(3) The first installation of any pump or pumping equipment in any well

(4) Any repair, as defined in this act, to any water well or pump

Section 7. Existing Installations

No well or pump installation in existence on the effective date of this act shall be required to conform to the provisions of subsection (a) of section 6 of this act, or any rules or regulations adopted pursuant thereto; provided, however, that any well now or hereafter abandoned, including any well deemed to have been abandoned, as defined in this act, shall be brought into compliance with the requirements of this act and any applicable rules or regulations with respect to abandonment of wells; and further provided, that any well or pump installation supplying water that is determined by the Department to be a health hazard must comply with the provisions of this act and applicable rules and regulations within a reasonable time after notification of such determination has been given.

Section 8. Inspections

(a) The Department is authorized to inspect any water well, abandoned water well, or pump installation for any well. Duly authorized representatives of the Department may at reasonable times enter upon, and shall be given access to, any premises for the purpose of such inspection.

(b) Upon the basis of such inspections, if the Department finds applicable laws, rules, or regulations have not been complied with, or that a health hazard exists, the Department shall disapprove the well and/or pump installation. If disapproved, no well or pump installation shall thereafter be used until brought into compliance and any health hazard is eliminated.

(c) Any person aggrieved by the disapproval of a well or pump installation shall be afforded the opportunity of a hearing as provided in section 13 of this act.

Section 9. Licenses

Every person who wishes to engage in such business as a water well contractor or pump installation contractor, or both, shall obtain from the Department a license to conduct such business.

(a) The Department may adopt, and from time to time amend, rules and regulations governing applications for water well contractor licenses or pump installation contractor licenses, provided that the Department shall license, as a water well contractor or pump installation contractor, any person properly making application therefor, who is not less than twenty-one (21) years of age, is of good moral character, has knowledge of rules and regulations adopted under this act, and has had not less than two (2) years' experience in the work for which he is applying for a license; and provided further, that the Department shall prepare an examination that each such applicant must pass in order to qualify for such license.

(b) This section shall not apply to any person who performs labor or services at the direction and under the personal supervision of a licensed water well contractor or pump installation contractor.

(c) A county, municipality, or other political subdivision of the State engaged in well drilling or pump installing shall be licensed under this act, but shall be exempt from paying the license fees for the drilling or installing done by regular employees of, and with equipment owned by, the governmental entity.

(d) Any person who was engaged in the business of a water well contractor or pump installation contractor, or both, for a period of two (2) years immediately prior to (date of enactment) shall, upon application made within twelve (12) months of (date of enactment), accompanied by satisfactory proof that he was so engaged, and accompanied by payment of the required fees, be licensed as a water well contractor, pump installation contractor, or both, as provided in subsection (a) of this section, without fulfilling the requirement that he pass any examination prescribed pursuant thereto.

(e) Any person whose application for a license to engage in business as a water well contractor or pump installation contractor has been denied, may request, and shall be granted, a hearing in the county where such complainant has his place of business before an appropriate official of [insert the name of the hearing body designated in section 13 of this act].

(f) Licenses issued pursuant to this section are not transferable and shall expire on _____ of each year. A license may be renewed without examination for an ensuing year by making application not later than thirty (30) days after the expiration date and paying the applicable fee. Such application shall have the effect of extending the validity of the current license until a new license is received or the applicant is notified by the Department that it has refused to renew his license. After _____ of each year, a license will be renewed only upon application and payment of the applicable fee plus a penalty of $_____ .

(g) Whenever the Department determines that the holder of any license issued pursuant to this section has violated any provision of this act, or any rule or regulation adopted pursuant thereto, the Department is authorized to suspend or revoke any such license. Any order issued pursuant to this subsection shall be served upon the license holder pursuant to the provisions of subsection (a) of section 12 of this act. Any such order shall become effective _____ days after service thereof, unless a written petition requesting hearing, under the procedure provided in section 13, is filed sooner. Any person aggrieved by any order issued after such hearing may appeal therefrom in any court of competent jurisdiction as provided by the laws of this State.

(h) No application for a license issued pursuant to this section may be made within one (1) year after revocation thereof.

Section 10. Exemptions

(a) Where the Department finds that compliance with all requirements of this act would result in undue hardship, an exemption from any one or more such requirements may be granted by the Department to the extent necessary to ameliorate such undue hardship and to the extent such exemption can be granted without impairing the intent and purpose of this act.

(b) Nothing in this act shall prevent a person who has not obtained a license pursuant to section 9 of this act from constructing a well or installing a pump on his own or leased property intended for use only in a single family house that is his permanent residence, or intended for use only for farming purposes on his farm, and where the waters to be produced are not intended for use by the public or in any residence other than his own. Such person shall comply with all rules and regulations as to construction of wells and installation of pumps and pumping equipment adopted under this act.

Section 11. Fees

The following fees are required:

(a) A fee of $_____ shall accompany each application for permission required under section 6(a) of this act.

(b) A fee of $_____ shall accompany each application for a license required under section 9 of this act.

Section 12. Enforcement

(a) Whenever the Department has reasonable grounds for believing that there has been a violation of this act, or any rule or regulation adopted pursuant thereto, the Department shall give written notice to the person or persons alleged to be in violation. Such notice shall identify the provision of this act, or regulation issued hereunder, alleged to be violated and the facts alleged to constitute such violation.

(b) Such notice shall be served in the manner required by law for the service of process upon person in a civil action, and may be accompanied by an order of the Department requiring described remedial action, which, if taken within the time specified in such order, will effect compliance with the requirements of this act and regulations issued hereunder. Such order shall become final unless a request for hearing as provided in section 13 of this act is made within ____ days from the date of service of such order. In lieu of such order, the Department may require the person or persons named in such notice to appear at a hearing, at a time and place specified in the notice.

Section 13. Hearing

[Unless already prescribed in State law, this section should be used to specify procedures for administrative hearing.]

Section 14. Judicial Review

[Unless already prescribed in State law, this section should be used to specify procedures for judicial review.]

Section 15. Penalties

Any person who violates any provision of this act, or regulations issued hereunder, or order pursuant hereto, shall be subject to a penalty of $____. Every day, or any part thereof, in which such violation occurs shall constitute a separate violation.

Section 16. Conflict With Other Laws

The provisions of any law, or regulation of any municipality establishing standards affording greater protection to the public health or safety, shall prevail within the jurisdiction of such municipality over the provisions of this act and regulations adopted hereunder.

Section 17. Severability

[Insert severability clause.]

Section 18. Effective Date

[Insert effective date.]

Index

www.ingramcontent.com/pod-product-compliance
Lightning Source LLC
Chambersburg PA
CBHW031811190326
41518CB00006B/282